I0071806

# Bulletin of the Philosophical Society of Washington

# Also from Westphalia Press

### westphaliapress.org

# Bulletin of the Philosophical Society of Washington

# Volume IX

---

## For the Year of 1886

WESTPHALIA PRESS
An imprint of Policy Studies Organization

**Bulletin of the Philosophical Society of Washington: Volume IX**
All Rights Reserved © 2017 by Policy Studies Organization

**Westphalia Press**
An imprint of Policy Studies Organization
1527 New Hampshire Ave., NW
Washington, D.C. 20036
info@ipsonet.org

ISBN-13: 978-1-63391-578-7
ISBN-10: 1-63391-578-6

Daniel Gutierrez-Sandoval, Executive Director
PSO and Westphalia Press

Updated material and comments on this edition
can be found at the Westphalia Press website:
www.westphaliapress.org

# BULLETIN

of the

# PHILOSOPHICAL SOCIETY

OF

# WASHINGTON,

## VOL. IX.

Containing the Minutes of the Society and of the Mathematical
Section for the year.

PUBLISHED BY THE CO-OPERATION OF THE SMITHSONIAN INSTITUTION.

WASHINGTON:

1887.

Stereotyped and Printed

By JUDD & DETWEILER,

Washington, D. C.

# CONTENTS.

# BULLETIN

OF THE

# PHILOSOPHICAL SOCIETY OF WASHINGTON.

———

## CONSTITUTION, RULES,

LISTS OF

## OFFICERS AND MEMBERS,

AND REPORTS OF

## SECRETARIES AND TREASURER.

# CONSTITUTION

OF

## THE PHILOSOPHICAL SOCIETY OF WASHINGTON.

ARTICLE I. The name of this Society shall be THE PHILOSOPHICAL SOCIETY OF WASHINGTON.

ARTICLE II. The officers of the Society shall be a President, four Vice-Presidents, a Treasurer, and two Secretaries.

ARTICLE III. There shall be a General Committee, consisting of the officers of the Society and nine other members.

ARTICLE IV. The officers of the Society and the other members of the General Committee shall be elected annually by ballot; they shall hold office until their successors are elected, and shall have power to fill vacancies.

ARTICLE V. It shall be the duty of the General Committee to make rules for the government of the Society, and to transact all its business.

ARTICLE VI. This constitution shall not be amended except by a three-fourths vote of those present at an annual meeting for the election of officers, and after notice of the proposed change shall have been given in writing at a stated meeting of the Society at least four weeks previously.

# STANDING RULES

FOR THE GOVERNMENT OF THE

## PHILOSOPHICAL SOCIETY OF WASHINGTON.

1. The Stated Meetings of the Society shall be held at 8 o'clock P. M. on every alternate Saturday; the place of meeting to be designated by the General Committee.

2. Notice of the time and place of meeting shall be sent to each member by one of the Secretaries.

When necessary, Special Meetings may be called by the President.

3. The Annual Meeting for the election of officers shall be the last stated meeting in the month of December.

The order of proceedings (which shall be announced by the Chair) shall be as follows:

First, the reading of the minutes of the last Annual Meeting.

Second, the presentation of the annual reports of the Secretaries, including the announcement of the names of members elected since the last Annual Meeting.

Third, the presentation of the annual report of the Treasurer.

Fourth, the announcement of the names of members who, having complied with section 14 of the Standing Rules, are entitled to vote on the election of officers.

Fifth, the election of President.

Sixth, the election of four Vice-Presidents.

Seventh, the election of Treasurer.

Eighth, the election of two Secretaries.

Ninth, the election of nine members of the General Committee.

Tenth, the consideration of Amendments to the Constitution of the Society, if any such shall have been proposed in accordance with Article VI of the Constitution.

Eleventh, the reading of the rough minutes of the meeting.

4. Elections of officers are to be held as follows:

In each case nominations shall be made by means of an informal ballot, the result of which shall be announced by the Secretary; after which the first formal ballot shall be taken.

In the ballot for Vice-Presidents, Secretaries, and members of the General Committee, each voter shall write on one ballot as many names as there are officers to be elected, viz., four on the first ballot for Vice-Presidents, two on the first for Secretaries, and nine on the first for members of the General Committee; and on each subsequent ballot as many names as there are persons yet to be elected; and those persons who receive a majority of the votes cast shall be declared elected.

If in any case the informal ballot result in giving a majority for any one, it may be declared formal by a majority vote.

5. The Stated Meetings, with the exception of the Annual Meeting, shall be devoted to the consideration and discussion of scientific subjects.

The Stated Meeting next preceding the Annual Meeting shall be set apart for the delivery of the President's Annual Address.

6. Sections representing special branches of science may be formed by the General Committee upon the written recommendation of twenty members of the Society.

7. Persons interested in science, who are not residents of the District of Columbia, may be present at any meeting of the Society, except the Annual Meeting, upon invitation of a member.

8. On request of a member, the President or either of the Secretaries may, at his discretion, issue to any person a card of invitation to attend a specified meeting. Five cards of invitation to attend a meeting may be issued in blank to the reader of a paper at that meeting.

9. Invitations to attend during three months the meetings of the Society and participate in the discussion of papers, may, by a vote of nine members of the General Committee, be issued to persons nominated by two members.

10. Communications intended for publication under the auspices of the Society shall be submitted in writing to the General Committee for approval.

11. Any paper read before a Section may be repeated, either entire or by abstract, before a general meeting of the Society, if such repetition is recommended by the General Committee of the Society.

12. It is not permitted to report the proceedings of the Society or its Sections for publication, except by authority of the General Committee.

13. *New members may be proposed in writing by three members of the Society for election by the General Committee; but no person shall be admitted to the privileges of membership unless he signifies his acceptance thereof in writing, and pays his dues to the Treasurer, within two months after notification of his election.

14. Each member shall pay annually to the Treasurer the sum of five dollars, and no member whose dues are unpaid shall vote at the Annual Meeting for the election of officers, or be entitled to a copy of the Bulletin.

In the absence of the Treasurer, the Secretary is authorized to receive the dues of members.

The names of those two years in arrears shall be dropped from the list of members.

Notice of resignation of membership shall be given in writing to the General Committee through the President or one of the Secretaries.

15. The fiscal year shall terminate with the Annual Meeting.

16. Any member who is absent from the District of Columbia for more than twelve consecutive months may be excused from payment of dues during the period of his absence, in which case he will not be entitled to receive announcements of meetings or current numbers of the Bulletin.

17. Any member not in arrears may, by the payment of one hundred dollars at any one time, become a life member, and be relieved from all further annual dues and other assessments.

All moneys received in payment of life membership shall be invested as portions of a permanent fund, which shall be directed solely to the furtherance of such special scientific work as may be ordered by the General Committee.

---

*Amended Oct. 9, 1886.

# STANDING RULES

OF THE

## GENERAL COMMITTEE OF THE PHILOSOPHICAL SOCIETY OF WASHINGTON.

1. The President, Vice-Presidents, and Secretaries of the Society shall hold like offices in the General Committee.

2. The President shall have power to call special meetings of the Committee, and to appoint Sub-Committees.

3. The Sub-Committees shall prepare business for the General Committee, and perform such other duties as may be entrusted to them.

4. There shall be two Standing Sub-Committees; one on Communications for the Stated Meetings of the Society, and another on Publications.

5. The General Committee shall meet at half-past seven o'clock on the evening of each Stated Meeting, and by adjournment at other times.

6. For all purposes, except for the amendment of the Standing Rules of the Committee or of the Society, and the election of members, six members of the Committee shall constitute a quorum.

7. The names of proposed new members recommended in conformity with section 13 of the Standing Rules of the Society, may be presented at any meeting of the General Committee, but shall lie over for at least four weeks before final action, and the concurrence of twelve members of the Committee shall be necessary to election.

The Secretary of the General Committee shall keep a chronological register of the elections and acceptances of members.

8. These Standing Rules, and those for the government of the Society, shall be modified only with the consent of a majority of the members of the General Committee.

# RULES

FOR THE

# PUBLICATION OF THE BULLETIN

OF THE

## PHILOSOPHICAL SOCIETY OF WASHINGTON.

---

1. The President's Annual Address shall be published in full.

2. The annual reports of the Secretaries and of the Treasurer shall be published in full.

3. When directed by the General Committee, any communication may be published in full.

4. Abstracts of papers and remarks on the same will be published, when presented to the Secretary by the author in writing within two weeks of the evening of their delivery, and approved by the Committee on Publications. Brief abstracts prepared by one of the Secretaries and approved by the Committee on Publications may also be published.

5. If the author of any paper read before a Section of the Society desires its publication, either in full or by abstract, it shall be referred to a committee to be appointed as the Section may determine.

The report of this committee shall be forwarded to the Publication Committee by the Secretary of the Section, together with any action of the Section taken thereon.

6. Communications which have been published elsewhere, so as to be generally accessible, will appear in the Bulletin by title only, but with a reference to the place of publication, if made known in season to the Committee on Publications.

# OFFICERS

## OF THE

# PHILOSOPHICAL SOCIETY OF WASHINGTON

### ELECTED DECEMBER 19, 1885.

---

*President* _____J. S. BILLINGS.

*Vice-Presidents* _____WILLIAM HARKNESS.  GARRICK MALLERY.

  C. E. DUTTON.  J. E. HILGARD.*

*Treasurer* _____ ROBERT FLETCHER.

*Secretaries* _____ G. K. GILBERT.  MARCUS BAKER.

## MEMBERS AT LARGE OF THE GENERAL COMMITTEE.

H. H. BATES.  F. W. CLARKE.

W. H. DALL.  G. B. GOODE.

J. R. EASTMAN.  HENRY FARQUHAR.

T. C. MENDENHALL.†  H. M. PAUL.

C. V. RILEY.

---

## STANDING COMMITTEES.

*On Communications:*

WILLIAM HARKNESS, *Chairman.*  G. K. GILBERT.  MARCUS BAKER.

*On Publications:*

G. K. GILBERT, *Chairman.*  ROBERT FLETCHER.  MARCUS BAKER.

S. F. BAIRD.‡

* Resigned membership Nov. 20, 1886.
† Resigned Oct. 9, 1886; vacancy filled by election of W. B. Taylor, Oct. 23, 1886.
‡ As Secretary of the Smithsonian Institution.

# OFFICERS

## OF THE

# PHILOSOPHICAL SOCIETY OF WASHINGTON

### ELECTED DECEMBER 18, 1886.

---

*President* ------------ WILLIAM HARKNESS.

*Vice-Presidents* -- ----- GARRICK MALLERY.  C. E. DUTTON.

J. R. EASTMAN.  G. K. GILBERT.

*Treasurer* ----- ----- ROBERT FLETCHER.

*Secretaries* ---- ------ MARCUS BAKER.  J. H. KIDDER.

## MEMBERS AT LARGE OF THE GENERAL COMMITTEE.

H. H. BATES.  F. W. CLARKE.

W. H. DALL.  E. B. ELLIOTT.

G. B. GOODE.  C. V. RILEY.

H. M. PAUL.  W. C. WINLOCK.

R. S. WOODWARD.

---

## STANDING COMMITTEES.

*On Communications:*

J. R. EASTMAN, *Chairman.*  MARCUS BAKER.  J. H. KIDDER.

*On Publications:*

MARCUS BAKER, *Chairman.*  ROBERT FLETCHER.  J. H. KIDDER.

S. F. BAIRD.

# LIST OF MEMBERS

OF THE

## PHILOSOPHICAL SOCIETY OF WASHINGTON.

Corrected to December 18, 1886.

Names of gentlemen here indicated as resigned will be omitted from future lists.

| NAME. | ADDRESS AND RESIDENCE. | Year of admission. |
|---|---|---|
| ABBE, Prof. CLEVELAND | Army Signal Office. 2017 I st. N. W. | 1871 |
| ABERT, Mr. S.T.(Sylvanus Thayer) | 810 19th st. N. W. | 1875 |
| ADAMS, Mr. CHARLES FREDERICK | Civil Service Commission. 316 C st. N. W. | 1885 |
| ADAMS, Mr. HENRY | 1603 H st. N. W. | 1881 |
| ALDIS, Hon. A. O. (Asa Owen) | 1765 Mass. ave. N. W. | 1873 |
| ANTISELL, Dr. THOMAS (*Founder*) | Patent Office. 1311 Q st. N. W. | 1871 |
| AVERY, Mr. ROBERT S. (Robert Stanton) | 820 A st. S. E. | 1879 |
| BAIRD, Prof. SPENCER F. (Spencer Fullerton) (*Founder*) | Smithsonian Institution. 1445 Mass. ave. N. W. | 1871 |
| BAKER, Prof. FRANK | 1315 Corcoran st. N. W. | 1881 |
| BAKER, Mr. MARCUS | Geological Survey. 1125 17th st. N. W. | 1876 |
| BANCROFT, GEORGE | 1623 H st. N. W., or, in summer, Newport, R. I. | 1875 |
| BARUS, Dr. CARL | Geological Survey. | 1885 |
| BATES, Mr. HENRY H. (Henry Hobart) | Patent Office. The Portland. | 1871 |
| BATES, Dr. N. L. (Newton Lemuel) U. S. N. | Navy Department. 1233 17th st. N. W. | 1866 |
| BEAN, Dr. T. H. (Tarleton Hoffman) | National Museum. 1616 19th st. N. W. | 1884 |
| BEARDSLEE, Capt. L. A. (Lester Anthony) U. S. N. (*Absent*) | Little Falls, N. Y. | 1875 |
| BELL, Mr. A. GRAHAM (Alexander Graham) | Scott Circle ; 1500 Rhode Island ave. | 1879 |
| BELL, Dr. C. A. (Chichester Alexander) (*Absent*) | University College. London, England. | 1881 |
| BENÉT, Gen. S. V. (Stephen Vincent) U. S. A. (*Founder*) | Ordnance Office, War Dept. 1717 I st. N. W. | 1871 |

| NAME. | ADDRESS AND RESIDENCE. | Year of admission. |
|---|---|---|
| BESSELS, Dr. EMIL | Glenn Dale, Md. | 1875 |
| BEYER, Dr. H. G. (Henry Gustav) | Bureau Med. & Surg.,Navy Dept. 1207 Connecticut ave. | 1886 |
| BILLINGS, Dr. JOHN S.(John Shaw) U. S. A. (*Founder*) | Surg. General's Office, U. S. A. 3027 N st. N. W. | 1871 |
| BIRNEY, Gen. WILLIAM | 456 Louisiana ave. 1901 Harewood ave., Le Droit Park. | 1879 |
| BIRNIE, Capt. ROGERS, Jr.,U. S. A. | Office Chief of Ordnance,U. S. A. 1341 New Hampshire ave. | 1876 |
| BODFISH, Mr. S. H. (Sumner Homer) (*Absent*) | Geological Survey. 605 F st. N. W. | 1883 |
| BOWLES, Asst. Nav. Constr. FRANCIS T. (Francis Tiffany) U. S. N. (*Absent*) | Navy Department. | 1884 |
| BROWN, Prof. S. J. (Stimson Joseph) U. S. N. (*Absent*) | Naval Academy, Annapolis, Md. | 1884 |
| BROWNE, Dr. J. MILLS (John Mills) U. S. N. | Navy Department. The Portland. | 1883 |
| BRYAN, Dr. J. H. (Joseph Hammond) | 1644 Conecticut ave. | 1886 |
| BURGESS, Mr. E. S. (Edward Sandford) | High School. 1120 13th st. N. W. | 1883 |
| BURNETT, Dr. SWAN M. (Swan Moses) | 1734 K st. N. W. | 1879 |
| BUSEY, Dr. SAMUEL C. (Samuel Clagett) | 901 16th st. N. W. | 1874 |
| CASEY, Col. THOMAS LINCOLN, U. S. A. (*Founder: absent*) | Army Building, cor. Green and Houston sts., New York city. | 1871 |
| CAZIARC, Lieut. L. V. (Louis Vasmer) U. S. A. (*Absent*) | Little Rock Barracks, Little Rock, Ark. | 1882 |
| CHAMBERLIN, Prof. T. C. (Thomas Crowder) | Geological Survey. | 1883 |
| CHATARD, Dr. THOMAS M.(Thomas Marean) | Geological Survey. Cosmos Club. | 1885 |
| CHICKERING, Prof. J. W., Jr. (John White) | Deaf Mute College, Kendall Green. | 1874 |
| CHRISTIE, Mr. ALEX. S. (Alexander Smyth) | Coast and Geodetic Survey Office. 507 6th st. N. W. | 1880 |
| CLARK, Mr. E. (Edward) | Architect's Office, Capitol. 417 4th st. N. W. | 1877 |
| CLARKE, Prof. F. W. (Frank Wigglesworth) | Geological Survey. 1425 Q st. N. W. | 1874 |
| COFFIN, Prof. J. H. C. (John Huntington Crane) U. S. N. (*Founder*) | 1901 I st. N. W. | 1871 |
| COMSTOCK, Prof. J. H. (John Henry) (*Absent*) | Cornell University, Ithaca, N.Y. | 1880 |
| COUES, Prof. ELLIOTT | Smithsonian Institution. 1726 N st. N. W. | 1874 |
| CRAIG, Lieut. ROBERT, U. S. A. (*Absent*) | War Department. | 1873 |

| NAME. | ADDRESS AND RESIDENCE. | Year of admission. |
|---|---|---|
| CRAIG, Dr. THOMAS (*Absent*) | Johns Hopkins Univ., Baltimore, Md. | 1879 |
| CUMMINGS, Prof. G. J. (George Jotham) | Howard University. | 1886 |
| CURTICE, Mr. COOPER | Agricultural Department. | 1886 |
| CURTIS, Mr. GEO. E. (George Edward) | Army Signal Office. 1401 16th st. N. W. | 1884 |
| DALL, Mr. WM. H. (William Healey) (*Founder*) | Care Smithsonian Institution. 1119 12th st. N. W. | 1871 |
| DARTON, Mr. NELSON H. (Nelson Horatio) | Geological Survey. 1101 K st. N. W. | 1886 |
| DAVIS, Commander C. H. (Charles Henry) U. S. N. | Navy Department. 1705 Rhode Island ave. | 1880 |
| DEAN, Dr. R. C. (Richard Crain) U. S. N. (*Absent*) | Navy Department. 45 Lafayette Place, New York city. | 1872 |
| DE CAINDRY, Mr. WM. A. (William Augustin) | Commissary General's Office. 1713 H st. N. W. | 1881 |
| DE LAND, Mr. THEODORE L. (Theodore Louis) | Treasury Department. 115 7th st. N. E. | 1880 |
| DEWEY, Mr. FRED. P. (Frederic Perkins) | National Museum. Lanier Heights. | 1884 |
| DILLER, Mr. J. S. (Joseph Silas) | Geological Survey. 1804 16th st. N. W. | 1884 |
| DOOLITTLE, Mr. M. H. (Myrick Hascall) | Coast and Geodetic Survey Office. 1925 I st. N. W. | 1876 |
| DUNWOODY, Lt. H. H. C. (Henry Harrison Chase) U. S. A. (*Absent*) | War Department. | 1873 |
| DUTTON, Capt. C. E. (Clarence Edward) U. S. A. | Geological Survey. 2119 H st. N. W. | 1872 |
| EARLL, Mr. R. EDWARD (Robert Edward) | Smithsonian Institution. 1336 T st. N. W. | 1884 |
| EASTMAN, Prof. J. R. (John Robie) U. S. N. | Naval Observatory. 1905 I st. N. W. | 1871 |
| EIMBECK, Mr. WILLIAM | Coast and Geodetic Survey Office. | 1884 |
| ELDREDGE, Dr. STEWART (*Absent*) | Yokohama, Japan. | 1871 |
| ELLIOTT, Mr. E. B. (Ezekiel Brown) (*Founder*) | Gov't Actuary, Treas. Dept. 1210 G st. N. W. | 1871 |
| EMMONS, Mr. S. F. (Samuel Franklin) | Geological Survey. 1708 H st. N. W. | 1883 |
| ENDLICH, Dr. F. M. (Frederic Miller) (*Absent*) | Reading, Pa. | 1873 |
| EWING, Gen. HUGH (*Absent*) | Lancaster, Ohio. | 1874 |
| FARQUHAR, Mr. EDWARD | Patent Office Library. 1915 H st. N. W. | 1876 |
| FARQUHAR, Mr. HENRY | Coast and Geodetic Survey Office. Brooks Station, D. C. | 1881 |
| FERREL, Prof. WILLIAM | 1641 Broadway, Kansas City, Mo. | 1872 |

| NAME. | ADDRESS AND RESIDENCE. | Year of admission. |
|---|---|---|
| FLETCHER, Dr. ROBERT | Surgeon General's Office, U. S. A. The Portland. | 1873 |
| FLINT, Mr. A. S. (Albert Stowell) | Naval Observatory. 1330 Riggs st. | 1882 |
| FLINT, Dr. J. M. (James Milton) U. S. N. | Navy Department. U. S. S. Albatross. | 1881 |
| FRISTOE, Prof. EDWARD T | Columbian University, cor. 15th and H sts. N. W. | 1873 |
| GALLAUDET, President E. M. (Edward Miner) | Deaf Mute College, Kendall Green. | 1875 |
| GANNETT, Mr. HENRY | Geological Survey. 1881 Harewood ave., Le Droit Park. | 1874 |
| GILBERT, Mr. G. K. (Grove Karl) | Geological Survey. 1424 Corcoran st. | 1873 |
| GODDING, Dr. W. W. (William Whitney) | Government Hospital for the Insane. | 1879 |
| GOOCH, Dr. F. A. (Frank Austin) (*Absent*) | Yale College, New Haven, Conn. | 1885 |
| GOODE, Mr. G. BROWN (George Brown) | National Museum. Summit ave., Lanier Heights. | 1874 |
| GOODFELLOW, Mr. EDWARD | Coast and Geodetic Survey Office. 1324 19th st. N. W. | 1875 |
| GORDON, Prof. J. C. (Joseph Claybaugh) | Deaf Mute College, Kendall Green. | 1886 |
| GORE, Prof. J. H. (James Howard) | Columbian University. 1305 Q st. N. W. | 1880 |
| GRAVES, Mr. WALTER H. (Walter Hayden) (*Absent*) | Denver, Colorado. | 1878 |
| GREELY, Lieut. A. W. (Adolphus Washington) U. S. A. | Army Signal Office. 1914 G st. N. W. | 1880 |
| GREEN, Mr. BERNARD R. (Bernard Richardson) | Office of Building for State, War and Navy Depts. 1738 N st. N. W. | 1879 |
| GREEN, Commander F. M. (Francis Mathews) U. S. N. (*Absent*) | Navy Department. | 1875 |
| GREENE, Prof. B. F. (Benjamin Franklin) U. S. N. (*Founder: absent*) | West Lebanon, N. H. | 1871 |
| GREENE, FRANCIS V. (Francis Vinton) (*Absent*) | 280 Broadway, New York city. | 1875 |
| GREGORY, Dr. JOHN M. (John Milton) | 15 Grant Place. | 1884 |
| GUNNELL, FRANCIS M., M. D., U. S. N. | 600 20th st. N. W. | 1879 |
| HAINS, Col. PETER C. (Peter Conover) | Engineer's Office, Potomac Riv. Improvement, 2136 Pa. ave. 1824 Jefferson Place. | 1879 |
| HAINS, Mr. ROBERT P. (Robert Peter) | Patent Office. 1714 13th st. N. W. | 1885 |

| NAME. | ADDRESS AND RESIDENCE. | Year of admission. |
|---|---|---|
| HALL, Prof. ASAPH, U. S. N. (*Founder*) | Naval Observatory. 2715 N st. N. W. | 1871 |
| HALL, Mr. ASAPH, Jr., (*Absent*) | Yale College Observatory, New Haven, Conn. | 1884 |
| HALLOCK, Dr. WILLIAM | Geological Survey. | 1885 |
| HAMPSON, Mr. THOMAS | Geological Survey. 504 Maple ave., Le Droit Park. | 1885 |
| HARKNESS, Prof. WILLIAM, U. S. N. (*Founder*) | Naval Observatory. Cosmos Club. | 1871 |
| HASSLER, Dr. FERDINAND A. (Ferdinand Augustus) (*Absent*) | Santa Aña, Los Angeles Co., Cal: | 1880 |
| HAYDEN, Dr. F. V. (Ferdinand Vandeveer) (*Founder: absent*) | Geological Survey. 1805 Arch st., Phila., Pa. | 1871 |
| HAZEN, Prof. H. A. (Henry Allen) | P. O. Box 427. 1416 Corcoran st. | 1882 |
| HAZEN, Gen. W. B. (William Babcock) U. S. A. | Army Signal Office. 1601 K st. N. W. | 1881 |
| HEAP, Major D. P. (David Porter) | Light House Board, Treas. Dept. 1618 Rhode Island ave. | 1884 |
| HENSHAW, Mr. H. W. (Henry Wetherbee) | Bureau of Ethnology. 13 Iowa Circle. | 1874 |
| HILGARD, Mr. J. E. (Julius Erasmus) (*Founder: resigned*) | | 1871 |
| HILL, Mr. G. W. (George William) | Nautical Almanac Office. 314 Indiana ave. N. W. | 1879 |
| HILL, Mr. ROBERT T. (Robert Thomas) | National Museum. 1464 Rhode Island ave. | 1886 |
| HILLEBRAND, Dr. W. F. (William Francis) | Geological Survey. 506 Maple ave., Le Droit Park. | 1886 |
| HITCHCOCK, Mr. ROMYN | National Museum. Osaka, Japan. | 1884 |
| HODGKINS, Prof. H. L. (Howard Lincoln) | Columbian University. 627 N st. N. W. | 1885 |
| HOLDEN, Prest. EDWARD SINGLETON (*Absent*) | University of California, Berkeley, Cal. | 1873 |
| HOLMES, Mr. W. H. (William Henry) | Geological Survey. 1100 O st. N. W. | 1879 |
| HOWELL, Mr. EDWIN E. (Edwin Eugene) (*Absent*) | 48 Oxford st., Rochester, N. Y. | 1874 |
| IDDINGS, Mr. JOSEPH P. (Joseph Paxson) | Geological Survey. 1028 Vermont ave. | 1885 |
| IRVING, Prof. R. D. (Roland Duer) | Geological Survey. | 1886 |
| JAMES, Rev. OWEN (*Absent*) | Scranton, Pa. | 1880 |
| JENKINS, Rear Admiral THORNTON A. (Thornton Alexander) U. S. N. (*Founder*) | 2115 Penna. ave. N. W. | 1871 |
| JOHNSON, Mr. A. B. (Arnold Burges) | Light House Board, Treas. Dept. 501 Maple ave., Le Droit Park. | 1878 |

| NAME. | ADDRESS AND RESIDENCE. | Year of admission. |
|-------|------------------------|--------------------|
| JOHNSON, Dr. JOSEPH TABER | 926 17th st. N. W. | 1879 |
| JOHNSON, Mr. WILLARD D. (Willard Drake) (*Absent*) | Geological Survey. | 1884 |
| JOHNSTON, Dr. W. W. (William Waring) | 1603 K st. N. W. | 1873 |
| KAUFFMANN, Mr. S. H. (Samuel Hay) | 1000 M st. N. W. | 1884 |
| KEITH, Prof. R. (Reuel) | Nautical Almanac Office. 1017 M st. N. W. | 1871 |
| KENASTON, Prof. C. A. (Carlos Albert) | Howard University. | 1886 |
| KERR, Mr. MARK B. (Mark Brickell) | Geological Survey. 722 21st st. N. W. | 1884 |
| KIDDER, Dr. J. H. (Jerome Henry) | Smithsonian Institution. 1816 N st. N. W. | 1880 |
| KILBOURNE, Lieut. C. E. (Charles Evans) U. S. A. (*Absent*) | War Department. | 1880 |
| KING, Dr. A. F. A. (Albert Freeman Africanus) | 726 13th st. N. W. | 1875 |
| KNOX, Hon. JOHN JAY (*Absent*) | Prest. Nat. Bank Republic, New York city. | 1874 |
| KUMMELL, Mr. C. H. (Charles Hugo) | Coast and Geodetic Survey Office. 608 Q st. N. W. | 1882 |
| LAWRENCE, Mr. WILLIAM | Bellefontaine, Ohio. | 1884 |
| LAWVER, Dr. W. P. (Winfield Peter) | Mint Bureau, Treas. Dept. 1912 I st. N. W. | 1881 |
| LEE, Dr. WILLIAM | 2111 Penna. ave. N. W. 1821 I st. N. W. | 1874 |
| LEFAVOUR, Mr. EDWARD B. (Edward Brown) | Cambridge, Mass. | 1882 |
| LINCOLN, Dr. N. S. (Nathan Smith) | 1514 H st. N. W. | 1871 |
| LOOMIS, Mr. E. J. (Eben Jenks) | Nautical Almanac Office. 1413 Stoughton st. N. W. | 1880 |
| LULL, Capt. E. P. (Edward Phelps) U. S. N. (*Absent*) | Navy Yard, Pensacola, Fla. | 1875 |
| McADIE, Mr. A. G. (Alexander George) | Army Signal Office. "Morleigh", Anacostia, D. C. | 1886 |
| McDONALD, Col. MARSHALL | U. S. Fish Commission. 1515 R st. N. W. | 1886 |
| McGEE, Mr. W J | Geological Survey. 920 14th st. N. W. | 1883 |
| McGUIRE, Mr. FRED. B. (Frederick Bauders) | 1416 F st. N. W. 614 E st. N. W. | 1879 |
| McMURTRIE, Prof. WILLIAM (*Absent*) | University of Illinois, Champaign, Ill. | 1876 |
| MAHER, Mr. JAMES A. (James Arran) | Geological Survey. 21 E st. N. W. | 1884 |
| MALLERY, Col. GARRICK, U. S. A. | Bureau of Ethnology. 1323 N st. N. W. | 1875 |

| NAME. | ADDRESS AND RESIDENCE. | Year of admission. |
|---|---|---|
| MANN, Mr. B: PICKMAN (Benjamin Pickman) | Department of Agriculture.<br>1918 Sunderland Place, N.W. | 1885 |
| MARCOU, Mr. J. B. (John Belknap) | Geological Survey.<br>Cosmos Club. | 1884 |
| MARTIN, Mr. ARTEMAS | Coast and Geodetic Survey Office.<br>55 C st. S. E. | 1886 |
| MARVIN, Prof. C. F. (Charles Frederick) | Army Signal Office.<br>1736 13th st. N. W. | 1885 |
| MARVIN, Mr. Jos. B. (Joseph Badger) (Absent) | Internal Revenue Bureau. | 1878 |
| MASON, Prof. OTIS T. (Otis Tufton) | National Museum.<br>1305 Q st. N. W. | 1875 |
| MATTHEWS, Dr. W. (Washington) U. S. A. | Surg. General's Office, U. S. A. | 1884 |
| MEIGS, Gen. M. C. (Montgomery Cunningham) U. S. A. (Founder) | 1239 Vermont ave. | 1871 |
| MENDENHALL, Prof. T. C. (Thomas Corwin) (Absent) | Terre Haute, Ind. | 1885 |
| MERRIAM, Dr. C. HART (Clinton Hart) | Department of Agriculture.<br>1912 Sunderland Place, N.W. | 1886 |
| MERRILL, Mr. GEORGE P. (George Perkins) | National Museum.<br>1602 19th st. N. W. | 1884 |
| MITCHELL, Prof. HENRY | Coast and Geodetic Survey Office.<br>1331 L st. N. W. | 1886 |
| MORGAN, Dr. E. CARROLL (Ethelbert Carroll) | 918 E st. N. W. | 1883 |
| MOSER, Lt. J. F. (Jefferson Franklin) U. S. N. (Absent) | Coast and Geodetic Survey Office. | 1885 |
| MURDOCH, Mr. JOHN | National Museum.<br>1441 Chapin st., College Hill. | 1884 |
| NEWCOMB, Prof. SIMON, U. S. N. (Founder) | Navy Department.<br>941 M st. N. W. | 1871 |
| NICHOLS, Dr. CHARLES H. (Charles Henry) (Absent) | Bloomingdale Asylum, Boulevard and 117th st., New York, N. Y. | 1872 |
| NICHOLSON, Mr. W. L. (Walter Lamb) (Founder) | Topographer, P. O. Dept.<br>2109 G st. N. W. | 1871 |
| NORDHOFF, Mr. CHARLES | 1731 K st. N. W. | 1879 |
| NORRIS, Dr. BASIL, U. S. A. (Absent) | Vancouver Barracks, Wash. Ter. | 1884 |
| NOTT, Judge C. C. (Charles Cooper) | Court of Claims.<br>826 Connecticut ave. N. W. | 1885 |
| OGDEN, Mr. HERBERT G. (Herbert Gouverneur) | Coast and Geodetic Survey Office.<br>1324 19th st. N. W. | 1884 |
| OSBORNE, Mr. J. W. (John Walter) | 212 Delaware ave. N. E. | 1878 |
| PARKE, Gen. JOHN G. (John Grubb) U. S. A. (Founder) | Engineer Bureau, War Dept.<br>16 Lafayette Square. | 1871 |
| PARKER, Dr. PETER (Founder) | 2 Lafayette Square. | 1871 |
| PARRY, Dr. CHARLES C. (Charles Christopher) (Absent) | Davenport, Iowa. | 1871 |

| NAME. | ADDRESS AND RESIDENCE. | Year of admission. |
|---|---|---|
| PAUL, Mr. H. M. (Henry Martyn) | Naval Observatory. 109 1st st. N. E. | 1877 |
| PEALE, Dr. A. C. (Albert Charles) | Geological Survey. 1010 Mass. ave. N. W. | 1874 |
| POE, Gen. O. M. (Orlando Metcalfe) U. S. A. (Absent) | 84 West Congress st., Detroit, Mich. | 1873 |
| POINDEXTER, Mr. W. M. (William Mundy) | 1505 Pennsylvania ave. 1227 15th st. N. W. | 1884 |
| POPE, Dr. B. F. (Benjamin Franklin) U. S. A. | Surg. General's Office, U. S. A. | 1882 |
| POWELL, Major J. W. (John Wesley) | Geological Survey. 910 M st. N. W. | 1874 |
| PRENTISS, Dr. D. W. (Daniel Webster) | 1101 14th st. N. W. | 1880 |
| PRITCHETT, Prof. H. S. (Henry Smith) (Absent) | Director of Observatory, Wash. University, St. Louis, Mo. | 1879 |
| RATHBUN, Mr. RICHARD | Smithsonian Institution. 1622 Mass. ave. | 1882 |
| RAY, Lieut. P. H. (Patrick Henry) U. S. A. | Fort Gaston, Cal. | 1884 |
| RENSHAWE, Mr. JNO. H. (John Henry) | Geological Survey. 1512 Kingman Place. | 1883 |
| RICKSECKER, Mr. EUGENE | Geological Survey. 1323 Q st. N. W. | 1884 |
| RILEY, Dr. C. V. (Charles Valentine) | Agricultural Department, or National Museum. 1700 13th st. N. W. | 1878 |
| RITTER, Mr. W. F. McK. (William Francis McKnight) | Nautical Almanac Office. 16 Grant Place. | 1879 |
| ROBINSON, Mr. THOMAS | Howard University. 6th st. N. W., cor. Lincoln. | 1884 |
| ROGERS, Mr. JOSEPH A. (Joseph Addison) (Absent) | Naval Observatory. | 1872 |
| RUSSELL, Mr. ISRAEL C. (Israel Cook) | Geological Survey. | 1882 |
| RUSSELL, Mr. THOMAS | Army Signal Office. 1447 Corcoran st. N. W. | 1883 |
| SALMON, Dr. D. E. (Daniel Elmer) | Agricultural Department. 12 Iowa Circle. | 1883 |
| SAMPSON, Commander W. T. (William Thomas) U. S. N. (Absent) | Naval Academy, Annapolis, Md. | 1883 |
| SAVILLE, Mr. J. H. (James Hamilton) | 1419 F st. N. W. 1315 M st. N. W. | 1871 |
| SCHOTT, Mr. CHARLES A. (Charles Anthony) (Founder) | Coast and Geodetic Survey Office. 212 1st st. S. E. | 1871 |
| SHELLABARGER, Hon. SAMUEL | Room 31 Kellogg Building. 812 17th st. N. W. | 1875 |
| SHERMAN, Hon. JOHN | U. S. Senate. 1319 K st. N. W. | 1874 |
| SHUFELDT, Dr. R. W. (Robert Wilson) U. S. A. (Absent) | Surg. Gen'l's Office, U. S. A., or Box 144 Smithsonian Inst. | 1881 |

| NAME. | ADDRESS AND RESIDENCE. | Year of admission. |
|---|---|---|
| SIGSBEE, Commander C. D. (Charles Dwight) U. S. N. (Absent) | Navy Department. | 1879 |
| SKINNER, Dr. J. O. (John Oscar) U. S. A. | Surg. General's Office, U. S. A. 1529 O st. N. W. | 1883 |
| SMILEY, Mr. CHAS. W. (Charles Wesley) | U. S. Fish Commission. 943 Mass. ave. | 1882 |
| SMITH, Chf. Eng. DAVID, U. S. N. | Navy Department. | 1876 |
| SMITH, Mr. EDWIN | Coast and Geodetic Survey Office. 2024 Hillyer Place. | 1880 |
| SNELL, Mr. MERWIN M. (Merwin Marie) | National Museum. 715 Mt. Vernon Place. | 1886 |
| SPOFFORD, Mr. A. R. (Ainsworth Rand) | Library of Congress. 1621 Mass. ave. N. W. | 1872 |
| STEARNS, Mr. ROBERT E. C. (Robert Edwards Carter) | Smithsonian Institution. 1635 13th st. N. W. | 1884 |
| STONE, Prof. ORMOND (Absent) | Leander McCormick Observatory, University of Virginia, Va. | 1874 |
| TAYLOR, Mr. F. W. (Frederick William) (Absent) | Care Smithsonian Institution. | 1881 |
| TAYLOR, Mr. WILLIAM B. (William Bower) (Founder) | Smithsonian Institution. 306 C st. N. W. | 1871 |
| THOMPSON, Prof. A. H. (Almon Harris) | Geological Survey. | 1875 |
| THOMPSON, Mr. GILBERT | Geological Survey. 1448 Q st. N. W. | 1884 |
| TODD, Prof. DAVID P. (David Peck) (Absent) | Amherst College Observatory, Amherst, Mass. | 1878 |
| TONER, Dr. J. M., (Joseph Meredith) | 615 Louisiana ave. | 1873 |
| TRENHOLM, Hon. WILLIAM L. (William Lee.) | Controller of the Currency. 1812 N st. N. W. | 1886 |
| TRUE, Mr. FREDERICK W. (Frederick William) | National Museum. 1335 N st. N. W. | 1882 |
| UPTON, Mr. WM. W. (William Wirt) | 1416 F st. N. W. 1746 M st. N. W. | 1882 |
| UPTON, Prof. WINSLOW (Absent) | Brown University, Providence, R. I. | 1880 |
| WALCOTT, Mr. C. D. (Charles Doolittle) | Geological Survey; National Museum. | 1883 |
| WALDO, Prof. FRANK (Absent) | Army Signal Office, Fort Myer, Va. | 1881 |
| WALKER, Mr. FRANCIS A. (Francis Amasa) (Absent) | Massachusetts Institute of Technology, Boston, Mass. | 1872 |
| WALLING, Mr. HENRY F. (Henry Francis) (Absent) | U. S. Geological Survey, Cambridge, Mass. | 1883 |
| WARD, Mr. LESTER F. (Lester Frank) | Geological Survey. 1464 Rhode Island ave. | 1876 |

| NAME. | ADDRESS AND RESIDENCE. | Year of admission. |
|---|---|---|
| WEBSTER, Mr. ALBERT L. (Albert Lowry) (*Absent*) | 107 Drexel Building, Broad st., New York city; West New Brighton, Staten Island, N.Y. | 1882 |
| WEED, Mr. WALTER H. (Walter Harvey) | Geological Survey. "The Grammercy", Vermont ave. | 1885 |
| WELLING, Mr. JAMES C. (James Clarke) | 1302 Connecticut ave. | 1872 |
| WHEELER, Capt. GEO. M. (George Montague) U. S. E. | Lock Box 93. 930 16th st. N. W. | 1873 |
| WHITE, Dr. C. A. (Charles Abiathar) | Geological Survey. 312 Maple ave., Le Droit Park | 1876 |
| WHITE, Dr. C. H. (Charles Henry) U. S. N. | Museum of Hygiene, 1744 G st. N. W. | 1884 |
| WILLIS, Mr. BAILEY | Geological Survey. 1823 H st. N. W. | 1885 |
| WILSON, H. M. (Herbert Michael) | Geological Survey. | 1885 |
| WILSON, Mr. J. ORMOND (James Ormond) | 1439 Massachusetts ave. N. W. | 1873 |
| WINLOCK, Mr. WILLIAM C. (William Crawford) | Naval Observatory. 718 21st st. N. W. | 1880 |
| WOOD, Mr. JOSEPH (*Absent*) | Supt. Motive Power, Penn. Co., Fort Wayne, Ind. | 1875 |
| WOOD, Lt. W. M. (William Maxwell) U. S. N. (*Absent*) | Navy Department. | 1871 |
| WOODWARD, Mr. R. S. (Robert Simpson) | Geological Survey. 1804 Columbia Road. | 1883 |
| WORTMAN, Dr. J. L. (Jacob Lawson) | Army Medical Museum. 1711 13th st. N. W. | 1885 |
| WRIGHT, Mr. GEO. M. (George Mitchell) (*Absent*) | Akron, Ohio. | 1885 |
| YARROW, Dr. H. C. (Harry Crécy) | Surgeon General's Office, U. S. A. 814 17th st. N. W. | 1874 |
| YEATES, Mr. W. S. (William Smith) | Smithsonian Institution. 1403 6th st. N. W. | 1884 |
| ZIWET, Mr. ALEXANDER | Coast and Geodetic Survey Office. 140 C st. S. E. | 1885 |
| ZUMBROCK, Dr. A. (Anton) | | 1875 |

## LIST OF DECEASED MEMBERS.

| Name. | Admitted. |
|---|---|
| Benjamin Alvord | 1872 |
| Orville Elias Babcock | 1871 |
| Theodorus Bailey | 1873 |
| Joseph K. Barnes | Founder |
| Henry Wayne Blair | 1884 |
| Horace Capron | Founder |
| Salmon Portland Chase | Founder |
| Frederick Collins | 1879 |
| Benjamin Faneuil Craig | Founder |
| Charles Henry Crane | Founder |
| Josiah Curtis | 1874 |
| Richard Dominicus Cutts | 1871 |
| Charles Henry Davis | 1874 |
| Frederick William Dorr | 1874 |
| Alexander B. Dyer | Founder |
| Amos Beebe Eaton | Founder |
| Charles Ewing | 1874 |
| Elisha Foote | Founder |
| John Gray Foster | 1873 |
| Leonard Dunnell Gale | 1874 |
| Isaiah Hanscom | 1873 |
| Joseph Henry | Founder |
| Franklin Benjamin Hough | 1879 |
| Andrew Atkinson Humphreys | Founder |
| Ferdinand Kampf | 1875 |
| Washington Caruthers Kerr | 1883 |
| Jonathan Homer Lane | Founder |
| Oscar A. Mack | 1872 |
| Archibald Robertson Marvine | 1874 |
| Fielding Bradford Meek | Founder |
| James William Milner | 1874 |
| Albert J. Myer | Founder |
| George Alexander Otis | Founder |
| Carlile Pollock Patterson | 1871 |
| Titian Ramsay Peale | Founder |
| Benjamin Peirce | Founder |
| John Campbell Riley | 1877 |
| John Rodgers | 1872 |
| Benjamin Franklin Sands | Founder |
| George Christian Shaeffer | Founder |
| Henry Robinson Searle | 1877 |
| William J. Twining | 1878 |
| Joseph Janvier Woodward | Founder |
| John Maynard Woodworth | 1874 |
| Mordecai Yarnall | 1871 |

## SUMMARY.

| | |
|---|---|
| Active members | 183 |
| Absent members | 54 |
| Total | 237 |
| Deceased members | 45 |

## CALENDAR FOR THE USE OF THE PHILOSOPHICAL SOCIETY,

*Showing the alternate SATURDAYS for holding Meetings during the several "Seasons" from 1884-'85 to 1907-'08, inclusive.*

PREPARED BY MR. E. B. ELLIOTT.

Submitted to the General Committee June 7, 1884, and ordered published.

| Years | October | November | December | Years | January | February | March | April | May | June |
|---|---|---|---|---|---|---|---|---|---|---|
| 1884 | 11, 25 | 8, 22 | 6, 20 | 1885 | 3, 17, 31 | 14, 28 | 14, 28 | 11, 25 | 9, 23 | 6, 20 |
| 1885 | 10, 24 | 7, 21 | 5, 19 | 1886 | 2, 16, 30 | 13, 27 | 13, 27 | 10, 24 | 8, 22 | 5, 19 |
| 1886 | 9, 23 | 6, 20 | 4, 18 | 1887 | 15, 29 | 12, 26 | 12, 26 | 9, 23 | 7, 21 | 4, 18 |
| 1887 | 15, 29 | 12, 26 | 10, 24 | 1888 | 7, 21 | 4, 18 | 3, 17, 31 | 14, 28 | 12, 26 | 9, 23 |
| 1888 | 13, 27 | 10, 24 | 8, 22 | 1889 | 5, 19 | 2, 16 | 2, 16, 30 | 13, 27 | 11, 25 | 8, 22 |
| 1889 | 12, 26 | 9, 23 | 7, 21 | 1890 | 4, 18 | 1, 15 | 1, 15, 29 | 12, 26 | 10, 24 | 7, 21 |
| 1890 | 11, 25 | 8, 22 | 6, 20 | 1891 | 3, 17, 31 | 14, 28 | 14, 28 | 11, 25 | 9, 23 | 6, 20 |
| 1891 | 10, 24 | 7, 21 | 5, 19 | 1892 | 2, 16, 30 | 13, 27 | 12, 26 | 9, 23 | 7, 21 | 4, 18 |
| 1892 | 15, 29 | 12, 26 | 10, 24 | 1893 | 7, 21 | 4, 18 | 4, 18 | 1, 15, 29 | 13, 27 | 10, 24 |
| 1893 | 14, 28 | 11, 25 | 9, 23 | 1894 | 6, 20 | 3, 17 | 3, 17, 31 | 14, 28 | 12, 26 | 9, 23 |
| 1894 | 13, 27 | 10, 24 | 8, 22 | 1895 | 5, 19 | 2, 16 | 2, 16, 30 | 13, 27 | 11, 25 | 8, 22 |
| 1895 | 12, 26 | 9, 23 | 7, 21 | 1896 | 4, 18 | 1, 15, 29 | 14, 28 | 11, 25 | 9, 23 | 6, 20 |
| 1896 | 10, 24 | 7, 21 | 5, 19 | 1897 | 2, 16, 30 | 13, 27 | 13, 27 | 10, 24 | 8, 22 | 5, 19 |
| 1897 | 9, 23 | 6, 20 | 4, 18 | 1898 | 15, 29 | 12, 26 | 12, 26 | 9, 23 | 7, 21 | 4, 18 |
| 1898 | 15, 29 | 12, 26 | 10, 24 | 1899 | 7, 21 | 4, 18 | 4, 18 | 1, 15, 29 | 13, 27 | 10, 24 |
| 1899 | 14, 28 | 11, 25 | 9, 23 | 1900 | 6, 20 | 3, 17 | 3, 17, 31 | 14, 28 | 12, 26 | 9, 23 |
| 1900 | 13, 27 | 10, 24 | 8, 22 | 1901 | 5, 19 | 2, 16 | 2, 16, 30 | 13, 27 | 11, 25 | 8, 22 |
| 1901 | 12, 26 | 9, 23 | 7, 21 | 1902 | 4, 18 | 1, 15 | 1, 15, 29 | 12, 26 | 10, 24 | 7, 21 |
| 1902 | 11, 25 | 8, 22 | 6, 20 | 1903 | 3, 17, 31 | 14, 28 | 14, 28 | 11, 25 | 9, 23 | 6, 20 |
| 1903 | 10, 24 | 7, 21 | 5, 19 | 1904 | 2, 16, 30 | 13, 27 | 12, 26 | 9, 23 | 7, 21 | 4, 18 |
| 1904 | 15, 29 | 12, 26 | 10, 24 | 1905 | 7, 21 | 4, 18 | 4, 18 | 1, 15, 29 | 13, 27 | 10, 24 |
| 1905 | 14, 28 | 11, 25 | 9, 23 | 1906 | 6, 20 | 3, 17 | 3, 17, 31 | 14, 28 | 12, 26 | 9, 23 |
| 1906 | 13, 27 | 10, 24 | 8, 22 | 1907 | 5, 19 | 2, 16 | 2, 16, 30 | 13, 27 | 11, 25 | 8, 22 |
| 1907 | 12, 26 | 9, 23 | 7, 21 | 1908 | 4, 18 | 1, 15, 29 | 14, 28 | 11, 25 | 9, 23 | 6, 20 |

## ANNUAL REPORT OF THE SECRETARIES.

### WASHINGTON, D. C., *December* 18, 1886.

*To the Philosophical Society of Washington:*

We have the honor to present the following statistical data for 1886:

The last Annual Report brought the record of membership down to January 16, 1886. The number of active members was then . . . . . . . . .  179

This number has been increased by the addition of 18 new members, by the return of 1 absent member, and by the reinstatement of 2 members previously dropped. It has been diminished by the departure of 9 members, by the resignation of 1, and by the dropping of 7 for non-payment of dues. There has been no death. The net increase of active members has thus been . . . .  4

And the active membership is now . . . . .  183

The roll of new members is:

| | | |
|---|---|---|
| N. L. BATES. | J. C. GORDON. | MARSHALL McDONALD. |
| H. G. BEYER. | R. T. HILL. | ARTEMAS MARTIN. |
| J. H. BRYAN. | W. F. HILLEBRAND. | C. H. MERRIAM. |
| G. J. CUMMINGS. | R. D. IRVING. | HENRY MITCHELL. |
| COOPER CURTICE. | C. A. KENASTON. | M. M. SNELL. |
| N. H. DARTON. | A. G. McADIE. | W. L. TRENHOLM. |

There have been 14 meetings for the presentation and discussion of papers (not including the public meeting of December 4); the average attendance has been 47. There have been 2 meetings of the Mathematical Section; average attendance 16.

In the general meetings 39 communications have been presented; in the mathematical 3. Altogether 42 communications have been made by 29 members and candidates for membership and by 2 guests. The number of members who have participated in the discussions is 38. The total number who have contributed to the scientific proceedings is 49, or 27 per cent. of the present active membership.

The General Committee has held 15 meetings; average attendance 12, the smallest attendance at any meeting being 7 and the largest 15.

G. K. GILBERT,
MARCUS BAKER,
*Secretaries.*

## THE REPORT OF THE TREASURER.

*Mr. President and Gentlemen:*

The report which I shall presently have the honor to submit to you shows the total receipts and disbursements for the fiscal year ending with this meeting.

The actual income belonging to the year 1886 was $878.00, and the expenditures for the same period were $451.50, leaving a net surplus of $426.50.

The unpaid dues of former years which have been collected this year, amount to $175.00.

By a resolution of the General Committee, passed May 22, 1886, the Treasurer was authorized to invest six hundred dollars of the surplus funds of the Society in six second-mortgage bonds of the Cosmos Club of this city. The present high premium on Government bonds reduces the annual interest upon them to about $2\frac{3}{4}$ per cent. The bonds purchased were obtained at par, and pay an interest of five per cent. per annum, the security being an extremely valuable piece of city property.

The assets of the Society consist of:

| | |
|---|---|
| 2 Government bonds, $1,000 and $500, at 4 per cent., | $1,500 00 |
| 1 " bond, 1,000, " 4½ " | 1,000 00 |
| 6 Cosmos Club bonds, " 5 " | 600 00 |
| Cash with Riggs & Co. . . . . . . . . . | 485 52 |
| Unpaid dues . . . . . . . . . | 265 00 |
| Total . $3,850 52 |

Of the "unpaid dues" it is probable that a part cannot be collected; on the other hand, the market value of the bonds is in excess of their face value.

Volume VIII of the Bulletin was duly sent in February to all members entitled to receive it, and to the societies and scientific journals with which it is the custom of the Philosophical Society to exchange its publications.

Dr.    *The Treasurer in Account with The Philosophical Society of Washington.*    Cr.

| 1886. | | | 1886. | | |
|---|---|---|---|---|---|
| | To balance cash on hand, Dec. 19, 1885 | $484 02 | Mar. 19. | By cash paid Judd & Detweiler for printing, binding, and wrapping Vol. VIII of the Bulletin | $284 53 |
| Dec. 18. | " cash received from sales of Bulletin | 8 00 | May 27. | By cash paid janitor for attendance at 15 meetings | 15 00 |
| | "    "    for dues of 1883 | 5 00 | | By cash paid for 6 bonds, Cosmos Club | 600 00 |
| | "    "    "    1884 | 45 00 | Dec. 15. | By cash paid for miscellaneous printing—circulars, postal cards, etc. | 58 45 |
| | "    "    "    1885 | 125 00 | | By cash paid for miscellaneous expenses of Secretaries and Treasurer for postage, stationery, clerical hire, etc. | 93 52 |
| | "    "    "    1886 | 750 00 | | Balance with Riggs & Co. | 485 52 |
| | "    "    for interest on bonds: | | | | |
| | On $1,500 at 4 per cent. ----$60 00 | | | | |
| | "   1,000  4½  "   ----  45 00 | | | | |
| | "    600   5   "   6 mo's interest---- 15 00 | | | | |
| | | 120 00 | | | |
| | | $1,537 02 | | | $1,537 02 |

ROBERT FLETCHER, *Treasurer.*

WASHINGTON, *December 18, 1886.*

# BULLETIN

OF THE

# PHILOSOPHICAL SOCIETY OF WASHINGTON.

---

## ANNUAL ADDRESS OF THE PRESIDENT.

# ANNUAL ADDRESS OF THE PRESIDENT,

## John S. Billings.

*Delivered December 4, 1886.*

---

### SCIENTIFIC MEN AND THEIR DUTIES.

---

*Mr. Chairman and Fellow-Members of the Philosophical Society:*

The honor of the presidency of such a society as this—carrying with it, as it does, the duty of giving at the close of the term of office an address on some subject of general interest, has been aptly compared to the little book mentioned in the Revelations of St. John—the little book which was "sweet in the mouth but bitter in the belly." I can only thank you for the honor, and ask your indulgence as to the somewhat discursive remarks which I am about to inflict upon you.

There is a Spanish proverb to the effect that no man can at the same time ring the bell and walk in the procession. For a few moments to-night I am to ring the bell, and being thus out of the procession I can glance for a moment at that part of it which is nearest. At first sight it does not appear to be a very homogeneous or well-ordered parade, for the individual members seem to be scattering in every direction, and even sometimes to be pulling in opposite ways; yet there is, after all, a definite movement of the whole mass in the direction of what we call progress. It is not this general movement that I shall speak of, but rather of the tendencies of individuals or of certain classes; some of the molecular movements, so to speak, which are not only curious and interesting of themselves, but which have an important bearing upon the mass, and some comprehension of which is necessary to a right understanding of the present condition and future prospects of science in this country.

The part of the procession of which I speak is made up of that body or class of men who are known to the public generally as "scientists," "scientific men," or "men of science." As commonly used, all these terms have much the same significance; but there are, nevertheless, shades of distinction between them, and in fact we need several other terms for purposes of classification of the rather heterogeneous mass to which they are applied. The word "scientist" is a coinage of the newspaper reporter, and, as ordinarily used, is very comprehensive. Webster defines a scientist as being "one learned in science, a savant"—that is, a wise man—and the word is often used in this sense. But the suggestion which the word conveys to my mind is rather that of one whom the public suppose to be a wise man, whether he is so or not, of one who claims to be scientific. I shall, therefore, use the term "scientist" in the broadest sense, as including scientific men, whether they claim to be such or not, and those who claim to be scientific men whether they are so or not.

By a scientific man I mean a man who uses scientific method in the work to which he specially devotes himself; who possesses scientific knowledge,—not in all departments, but in certain special fields. By scientific knowledge we mean knowledge which is definite and which can be accurately expressed. It is true that this can rarely be done completely, so that each proposition shall precisely indicate its own conditions, but this is the ideal at which we aim. There is no man now living who can properly be termed a complete savant, or scientist, in Webster's sense of the word. There are a few men who are not only thoroughly scientific in their own special departments, but are also men possessed of much knowledge upon other subjects and who habitually think scientifically upon most matters to which they give consideration; but these men are the first to admit the incompleteness and superficiality of the knowledge of many subjects which they possess, and to embrace the opportunity which such a society as this affords of meeting with students of other branches and of making that specially advantageous exchange in which each gives and receives, yet retains all that he had at first.

Almost all men suppose that they think scientifically upon all subjects; but, as a matter of fact, the number of persons who are so free from personal equation due to heredity, to early associations, to emotions of various kinds, or to temporary disorder of the digestive or nervous machinery that their mental vision is at all times achromatic and not astigmatic, is very small indeed.

Every educated, healthy man possesses some scientific knowledge, and it is not possible to fix any single test or characteristic which will distinguish the scientific from the unscientific man. There are scientific tailors, bankers, and politicians, as well as physicists,

chemists, and biologists. Kant's rule, that in each special branch
of knowledge the amount of science, properly so called, is equal to
the amount of mathematics it contains, corresponds to the definition
of pure science as including mathematics and logic, and nothing
else. It also corresponds to the distinction which most persons,
consciously or unconsciously, make between the so-called physical,
and the natural or biological sciences. Most of us, I presume, have
for the higher mathematics, and for the astronomers and physicists
who use them, that profound respect which pertains to comparative
ignorance, and to a belief that capacity for the higher branches
of abstract analysis is a much rarer mental quality than are those
required for the average work of the naturalist. I do not, however,
propose to discuss the hierarchy of the sciences; and the term science
is now so generally used in the sense of knowledge, more or less
accurate, of any subject, more especially in the relations of causes
and effects, that we must use the word in this sense, and leave to
the future the task of devising terms which will distinguish the
sciences, properly so called, from those branches of study and occu-
pation of which the most that can be said is that they have a scientific side. It is a sad thing that words should thus become polar-
ized and spoiled, but there seems to be no way of preventing it.

In a general way we may say that a scientific man exercises the
intellectual more than the emotional faculties, and is governed by
his reason rather than by his feelings. He should be a man of both
general and special culture, who has a little accurate information
on many subjects and much accurate information on some one or
two subjects, and who, moreover, is aware of his own ignorance and
is not ashamed to confess it.

We must admit that many persons who are known as scientists
do not correspond to this definition. Have you never heard, and
perhaps assented to, some such statements as these: "Smith is a
scientist, but he doesn't seem to have good, common sense," or " he
is a scientific crank?"

The unscientific mind has been defined as one which "is willing
to accept and make statements of which it has no clear conceptions
to begin with, and of whose truth it is not assured. It is the state
of mind where opinions are given and accepted without ever being
subjected to rigid tests." Accepting this definition, and also the
implied definition of a scientific mind as being the reverse of this,
let us for a moment depart from the beaten track which presi-

dential addresses usually follow, and instead of proceeding at once to eulogize the scientific mind and to recapitulate the wonderful results it has produced, let us consider the unscientific mind a little, not in a spirit of lofty condescension and ill-disguised contempt, but sympathetically, and from the best side that we can find. As this is the kind of mind which most of us share with our neighbors, to a greater or less degree, it may be as well not to take too gloomy a view of it. In the first place, the men with unscientific minds form the immense majority of the human race.

Our associations, habits, customs, laws, occupations, and pleasures are, in the main, suited to these unscientific minds; whose enjoyment of social intercourse, of the every-day occurrences of life, of fiction, of art, poetry, and the drama is, perhaps, none the less because they give and accept opinions without subjecting them to rigid tests. It is because there are a goodly number of men who do this that the sermons of clergymen, the advice of lawyers, and the prescriptions of physicians have a market value. This unscientific public has its uses. We can at least claim that we furnish the materials for the truly scientific mind to work with and upon; it is out of this undifferentiated mass that the scientific mind supposes itself to be developed by specialization, and from it that it obtains the means of its own existence. The man with the unscientific mind, who amuses himself with business enterprises, and who does not care in the least about ohms or pangenesis, may, nevertheless, be a man who does as much good in the world, is as valuable a citizen, and as pleasant a companion as some of the men of scientific minds with whom we are acquainted.

And in this connection I venture to express my sympathy for two classes of men who have in all ages been generally condemned and scorned by others, namely, rich men and those who want to be rich.

I do not know that they need the sympathy, for our wealthy citizens appear to support with much equanimity the disapprobation with which they are visited by lecturers and writers—a condemnation which seems in all ages to have been bestowed on those who have by those who have not.

So far as those who actually are rich are concerned, we may, I suppose, admit that a few of them—those who furnish the money to endow universities and professorships, to build laboratories, or to furnish in other ways the means of support to scientific men—are not wholly bad. Then, also, it is not always a man's own fault that l.e is rich; even a scientist may accidentally and against his will become rich.

As to those who are not rich, but who wish to be rich, whose chief desire and object is to make money, either to avoid the necessity for further labor, or to secure their wives and children from want, or for the sake of power and desire to rule, I presume it is unsafe to try to offer any apologies for their existence. But when it is claimed for any class of men, scientists or others, that they do not want these things it is well to remember the remarks made by old Sandy Mackay after he had heard a sermon on universal brotherhood: "And so the deevil's dead. Puir auld Nickie; and him so little appreciated, too. Every gowk laying his sins on auld Nick's back. But I'd no bury him until he began to smell a wee strong like. It's a grewsome thing is premature interment."

I have tried to indicate briefly the sense in which the terms "scientist" and "scientific man" are to be used and understood, and you see it is not an easy matter. The difficulty is less as regards the term "man of science." By this expression we mean a man who belongs to science peculiarly and especially, whose chief object in life is scientific investigation, whose thoughts and hopes and desires are mainly concentrated upon his search for new knowledge, whose thirst for fresh and accurate information is constant and insatiable. These are the men who have most advanced science, and whom we delight to honor, more especially in these later days, by glowing eulogiums of their zeal, energy, and disinterestedness.

The man of science, as defined by his eulogists, is the *beau idéal* of a philosopher, a man whose life is dedicated to the advancement of knowledge for its own sake, and not for the sake of money or fame, or of professional position or advancement. He undertakes scientific investigations exclusively or mainly because he loves the work itself, and not with any reference to the probable utility of the results. Such men delight in mental effort, or in the observation of natural phenomena, or in experimental work, or in historical research, in giving play to their imagination, in framing hypotheses and then in endeavoring to verify or disprove them, but always the main incentive is their own personal satisfaction (with which may be mingled some desire for personal fame), and not the pleasure or the good of others. Carried to an extreme, the eulogy of such men and their work is expressed in the toast of the Mathematical Society of England: "Pure mathematics; may it never be of use to any man!" Now, it is one thing to seek one's own pleasure, and quite another thing to pride one's self upon doing so. The men who do their scientific work for the love of it do some of the best work, and, as a rule, do not pride themselves on it, or feel or express contempt for those who seek their pleasure and amusement in other directions. It is only from a certain class of eulogists of pure science, so called, that we get such specimens of scientific "dudeism" as the toast just quoted, opposed to which may be cited the Arab saying that "A wise man without works is like a cloud without water."

There are other men who devote themselves to scientific work, but who prefer to seek information that may be useful; who try to advance our knowledge of Nature's laws in order that man may know how to adapt himself and his surroundings to those laws, and thus be healthier and happier. They make investigations, like the men of pure science—investigations in which they may or may not take pleasure, but which they make, even if tedious and disagreeable, for the sake of solving some problem of practical importance. These are the men who receive from the public the most honor, for it is seen that their work benefits others. After all, this is not peculiar to the votaries of science. In all countries and all times, and among all sorts and conditions of men, it has always been agreed that the best life, that which most deserves praise, is that which is devoted to the helping others, which is unselfish, not stained by envy or jealousy, and which has as its main pleasure and spring of action the desire of making other lives more pleasant, of bringing light into the dark places, of helping humanity.

But, on the other hand, the man who makes a profession of doing this, and who makes a living by so doing, the professional philanthropist, whether he be scientist or emotionalist, is by no means to be judged by his own assertions. Some wise German long ago remarked that "*Esel singen schlecht, weil sie zu hoch anstimmen*"—that is, "asses sing badly because they pitch their voices too high," and it is a criticism which it is well to bear in mind.

In one of the sermons of Kin O* the preacher tells the story of a powerful clam who laughed at the fears of other fish, saying that when he shut himself up he felt no anxiety; but on trying this method on one occasion when he again opened his shell he found himself in a fishmonger's shop. And to rely on one's own talents,

---

* Cornhill Magazine, August, 1869, p. 196.

on the services one may have rendered, on cleverness, judgments strength, or official position, and to feel secure in these, is to court the fate of the clam.

There are not very many men of science, and there are no satisfactory means of increasing the number; it is just as useless to exhort men to love science, or to sneer at them because they do not, as it is to advise them to be six feet three inches high or to condemn a man because his hair is not red.

While the ideal man of science must have a "clear, cold, keen intellect, as inevitable and as merciless in its conclusions as a logic engine," it would seem that, in the opinion of some, his greatness and superiority consists not so much in the amount of knowledge he possesses, or in what he does with it, as in the intensity and purity of his desire for knowledge.

This so-called thirst for knowledge must be closely analogous to an instinctive desire for exercise of an organ or faculty, such as that which leads a rat to gnaw, or a man of fine physique to delight in exercise. Such instincts should not be neglected. If the rat does not gnaw, his teeth will become inconvenient or injurious to himself, but it is not clear that he deserves any special eulogium merely because he gnaws.

It will be observed that the definition of a scientific man or man of science, says nothing about his manners or morals. We may infer that a man devoted to science would have neither time nor inclination for dissipation or vice; that he would be virtuous either because of being passionless or because of his clear foresight of the consequences of yielding to temptation.

My own experience, however, would indicate that either this inference is not correct or that some supposed scientific men have been wrongly classified as such. How far the possession of a scientific mind and of scientific knowledge compensates, or atones for, ill-breeding or immorality, for surliness, vanity, and petty jealousy, for neglect of wife or children, for uncleanliness, physical and mental, is a question which can only be answered in each individual case; but the mere fact that a man desires knowledge for its own sake appears to me to have little to do with such questions. I would prefer to know whether the man's knowledge and work is of any use to his fellow-men, whether he is the cause of some happiness in others which would not exist without him. And it may be noted that while utility is of small account in the eyes of some eulogist, of the man of science they almost invariably base their claims for his honor and support upon his usefulness.

The precise limit beyond which a scientist should not make money has not yet been precisely determined, but in this vicinity there are some reasons for thinking that the maximum limit is about $5,000 per annum. If there are any members of the Philosophical Society of Washington who are making more than this, or who, as the result of careful and scientific introspection, discover in themselves the dawning of a desire to make more than this, they may console themselves with the reflection that the precise ethics and etiquette which should govern their action under such painful circumstances have not yet been formulated. The more they demonstrate their indifference to mere pecuniary considerations the more creditable it is to them; so much all are agreed upon; but this is nothing new, nor is it specially applicable to scientists. Yet while each may and must settle such questions as regards himself for himself, let him be very cautious and chary about trying to settle them for other people. Denunciations of other men engaged in scientific pursuits on the ground that their motives are not the proper ones are often based on insufficient or inaccurate knowledge, and seldom, I think, do good.

This is a country and an age of hurry, and there seems to be a desire to rush scientific work as well as other things. One might suppose, from some of the literature on the subject, that the great object is to make discoveries as fast as possible; to get all the mathematical problems worked out; all the chemical combinations made; all the insects and plants properly labeled; all the bones and muscles of every animal figured and described. From the point of view of the man of science there does not seem to be occasion for such haste. Suppose that every living thing were known, figured, and described. Would the naturalist be any happier? Those who wish to make use of the results of scientific investigation of course desire to hasten the work, and when they furnish the means we cannot object to their urgency. Moreover, there is certainly no occasion to fear that our stock of that peculiar form of bliss known as ignorance will be soon materially diminished.

From my individual point of view, one of the prominent features in the scientific procession is that part of it which is connected with Government work. Our Society brings together a large number of scientific men connected with the various Departments; some of them original investigators; most of them men whose chief, though not only, pleasure is study. A few of them have important administrative duties, and are brought into close relations with the heads of Departments and with Congress. Upon men in such positions a double demand is made, and they are subject to criticism from two very different standpoints. On the one hand are the scientists, calling for investigations which shall increase knowledge without special reference to utility, and sometimes asking that employment be given to a particular scientist on the ground that the work to which he wishes to devote himself is of no known use, and therefore will not support him. On the other hand is the demand from the business men's point of view—that they shall show practical results; that in demands for appropriations from the public funds they shall demonstrate that the use to be made of such appropriations is for the public good, and that their accounts shall show that the money has been properly expended—"properly," not merely in the sense of usefully, but also in the legal sense—in the sense which was meant by Congress in granting the funds. Nay, more, they must consider not only the intentions of Congress but the opinions of the accounting officers of the Treasury, the comptroller and auditor, and their clerks, and not rely solely on their own interpretation of the statutes, if they would work to the best advantage, and not have life made a perpetual burden and vexation of spirit.

There is a tendency on the part of business men and lawyers to the belief that scientific men are not good organizers or administrators, and should be kept in leading strings; that it is unwise to trust them with the expenditure of, or the accounting for, money, and that the precise direction in which they are to investigate should be pointed out to them. In other words, that they should be made problem-solving machines as far as possible.

When we reflect on the number of persons who, like Mark Twain's cat, feel that they are "nearly lightning on superintending;" on the desire for power and authority, which is almost universal, the tendency to this opinion is not to be wondered at. Moreover, as regards the man of science, there is some reason for it in the very terms by which he is defined, the characteristics for which he is chiefly eulogized.

The typical man of science is, in fact, in many cases an abnormity, just as a great poet, a great painter, or a great musician is

apt to be, and this not only in an unusual development of one part
of the brain, but in an inferior development in others.  True, there
are exceptions to this rule—great and illustrious exceptions ; but I
think we must admit that the man of science often lacks tact, and
is indifferent to and careless about matters which do not concern
his special work, and especially about matters of accounts and
pecuniary details.  If such a man is at the head of a bureau, whose
work requires many subordinates and the disbursement of large
sums of money, he may consider the business management of his
office as a nuisance, and delegate as much of it as possible to some
subordinate official, who, after a time, becomes the real head and
director of the bureau.  Evil results have, however, been very rare,
and the recognition of the possibility of their occurrence is by no
means an admission that they are a necessity, and still less of the
proposition that administrative officers should not be scientific men.

I feel very sure that there are always available scientific men,
thoroughly well informed in their several departments, who are also
thoroughly good business men, and are as well qualified for administrative work as any.  When such men are really wanted they can
always be found, and, as a matter of fact, a goodly number of them
have been found, and are now in the Government service.

The head of a bureau has great responsibilities; and while his
position is, in many respects, a desirable one, it would not be eagerly
sought for by most scientific men if its duties were fully understood.

In the first place the bureau chief must give up a great part of
his time to routine hack work.  During his business, or office, hours
he can do little else than this routine work, partly because of its
amount, and partly because of the frequent interruptions to which
he is subjected.  His visitors are of all kinds and come from all
sorts of motives—some to pass away half an hour, some to get information, some seeking office.  It will not work well if he takes the
ground that his time is too important to be wasted on casual callers
and refers them to some assistant.

In the second place he must, to a great extent at least, give up the
pleasure of personal investigation of questions that specially interest
him, and turn them over to others.  It rarely happens that he can
carry out his own plans in his own way, and perhaps it is well that
this should be the case.  The general character of his work is usually
determined for him either by his predecessors, or by Congress, or by
the general consensus of opinion of scientific men interested in the

particular subject or subjects to which it relates. This last has very properly much weight; in fact, it has much more weight than one might suppose, if he judged from some criticisms made upon the work of some of our bureaus whose work is more or less scientific. In these criticisms it is urged that the work has not been properly planned and correlated; that it should not be left within the power of one man to say what should be done; that the plans for work should be prepared by disinterested scientific men—as, for instance, by a committee of the National Academy—and that the function of the bureau official should be executive only.

I have seen a good deal of this kind of literature within the last ten or twelve years, and some of the authors of it are very distinguished men in scientific work; yet I venture to question the wisdom of such suggestions. As a rule, the plans for any extended scientific work to be undertaken by a Government department are the result of very extended consultations with specialists, and meet with the approval of the majority of them. Were it otherwise the difficulties in obtaining regular annual appropriations for such work would be great and cumulative, for in a short time the disapproval of the majority of the scientific public would make itself felt in Congress. It is true that the *vis inertiæ* of an established bureau is very great. The heads of Departments change with each new administration, but the heads of bureaus remain; and if an unfit man succeeds in obtaining one of these positions, it is a matter of great difficulty to displace him; but it seems to me to be wiser to direct the main effort to getting right men in right places rather than to attempt to elaborate a system which shall give good results with inferior men as the executive agents, which attempt is a waste of energy.

You are all familiar with the results of the inquiry which has been made by a Congressional committee into the organization and work of certain bureaus which are especially connected with scientific interests, and with the different opinions which this inquiry has brought out from scientific men. I think that the conclusion of the majority of the committee, that the work is, on the whole, being well done, and that the people are getting the worth of their money, is generally assented to. True, some mistakes have been made, some force has been wasted, some officials have not given satisfaction; but is it probable that any other system would give so much better results that it is wise to run the risks of change?

This question brings us to the only definite proposition which has been made in this direction, namely, the proposed Department of Science, to which all the bureaus whose work is mainly scientific, such as the Coast Survey, the Geological Survey, the Signal Service, the Naval Observatory, etc., shall be transferred.

The arguments in favor of this are familiar to you, and, as regards one or two of the bureaus, it is probable that the proposed change would effect an improvement; but as to the desirability of centralization and consolidation of scientific interests and scientific work into one department under a single head, I confess that I have serious doubts.

One of the strongest arguments in favor of such consolidation that I have seen is the address of the late president of the Chemical Society of Washington, Professor Clarke, "On the Relations of the Government to Chemistry," delivered about a year ago. Professor Clarke advises the creation of a large, completely-equipped laboratory, planned by chemists and managed by chemists, in which all the chemical researches required by any department of the Government shall be made, and the abandonment of individual laboratories in the several bureaus on the ground that these last are small, imperfectly equipped, and not properly specialized; that each chemist in them has too broad a range of duty and receives too small a salary to command the best professional ability. He would have a national laboratory, in which one specialist shall deal only with metals, another with food products, a third with drugs, etc., while over the whole, directing and correlating their work, shall preside the ideal chemist, the all-round man, recognized as the leader of the chemists of the United States. And so should the country get better and cheaper results. It is an enticing plan and one which might be extended to many other fields of work. Granting the premises that we shall have the best possible equipment, with the best possible man at the head of it, and a sufficient corps of trained specialists, each of whom will contentedly do his own work as directed and be satisfied, so that there shall be no jealousies, or strikes, or boycotting, and we have made a long stride toward Utopia. But before we centralize in this way we must settle the question of classification. Just as in arranging a large library there are many books which belong in several different sections, so it is in applied science. Is it certain that the examination of food products or of drugs should be made under the direction of the national chemist rather than under that of the Departments which are most interested in the composition and quality of these articles? This does not seem to me to be a self-evident proposition by any means.

The opinion of a scientific man as to whether the Government should or should not undertake to carry out any particular branch of scientific research and publish the results, whether it should attempt to do such work through officers of the Army and Navy, or more or less exclusively through persons specially employed for the purpose, whether the scientific work shall be done under the direction of those who wish to use, and care only for, the practical results, or whether the scientific man shall himself be the administrative head and direct the manner in which his results shall be applied; the opinion of a scientific man on such points, I say, will differ according to the part he expects or desires to take in the work, according to the nature of the work, according to whether he is an Army or Navy officer or not, according to whether he takes more pleasure in scientific investigations than in administrative problems, and so forth.

It is necessary, therefore, to apply a correction for personal equation to each individual set of opinions before its true weight and value can be estimated, and, unfortunately, no general formula for this purpose has yet been worked out.

I can only indicate my own opinions, which are those of an Army officer, who has all he wants to do, who does not covet any of his neighbors' work or goods, and who does not care to have any more masters than those whom he is at present trying to serve. You see that I give you some of the data for the formula by which you are to correct my statements, but this is all I can do.

I am not inclined at present to urge the creation of a department of science as an independent department of the Government having at its head a Cabinet officer. Whether such an organization may become expedient in the future seems to me doubtful; but at all events I think the time has not yet come for it.

I do not believe that Government should undertake scientific work merely or mainly because it is scientific, or because some useful results may possibly be obtained from it. It should do, or cause to be done, such scientific work as is needful for its own information and guidance when such work cannot be done, or cannot be done so cheaply or conveniently, by private enterprise. Some kinds of work it can best have done by private contract, and not by officials; others, by its own officers.   To this last class belong those branches of scientific investigation, or the means for promoting them, which require long-continued labor and expenditure on a uniform plan—such as the work of the Government Observatory, of the Government surveys, of the collection of the statistics which are so much needed for legislative guidance, and in which we are at present so deficient, the formation of museums and libraries, and so forth.

Considering the plans and operations of these Government institutions from the point of view of the scientific public, it is highly desirable that they should contribute to the advancement of abstract science, as well as to the special practical ends for which they have been instituted; but from the point of view of the legislator, who has the responsibility of granting the funds for their support, the practical results should receive the chief consideration, and therefore they should be the chief consideration on the part of those who are to administer these trusts.   It must be borne in mind that while the average legislator is, in many cases, not qualified to judge *a priori* as to what practical results may be expected from a given plan for scientific work, he is, nevertheless, the court which is to decide the question according to the best evidence which he can get, or, rather, which is brought before him, and it is no unimportant part of the duty of those who are experts in these matters to furnish such evidence.

But in saying that practical results should be the chief consideration of the Government and of its legislative and administrative agents it is not meant that these should be the only considerations. In the carrying out of any extensive piece of work which involves the collection of data, experimental inquiry, or the application of scientific results under new conditions there is more or less opportunity to increase knowledge at the same time and with comparatively little increased cost.   Such opportunity should be taken advantage of, and is also a proper subsidiary reason for adopting one plan of work in preference to another, or for selecting for appointment persons qualified not only to do the particular work which is the main object, but also for other allied work of a more purely scientific character.

On the same principle it seems to me proper and expedient that when permanent Government employees have at times not enough to do in their own departments, and can be usefully employed in scientific work, it is quite legitimate and proper to thus make use of them. For example, it is desirable that this country should have such an organization of its Army and Navy as will permit of rapid expansion when the necessity arises, and this requires that more officers shall be educated and kept in the service than are needed for military and naval duty in time of peace. It has been the policy of the Government to employ some of these officers in work connected with other departments, and especially in work which requires such special training, scientific or administrative, or both, as such officers possess. To this objections are raised, which may be summed up as follows:

First, that such officers ought not to be given positions which would otherwise be filled by civilian scientists, because these places are more needed by the civilians as a means of earning subsistence, and because it tends to increase the competition for places and to lower salaries. Put in other words, the argument is that it is injurious to the interests of scientific men, taken as a body, that the Government should employ in investigations or work requiring special knowledge and skill men who have been educated and trained at its expense, and who are permanently employed and paid by it. This is analogous to the trades union and the anti-convict labor platforms.

The second objection is that Army and Navy officers do not, as a rule, possess the scientific and technical knowledge to properly perform duties lying outside of the sphere of the work for which they have been educated, and that they employ as subordinates really skilled scientific men, who make the plans and do most of the work, but do not receive proper credit for it. The reply to this is that it is a question of fact in each particular case, and that if the officer is able to select and employ good men to prepare the plans and to do the work, this in itself is a very good reason for giving him the duty of such selection and employment.

A third objection is that when an officer of the Army or Navy is detailed for scientific or other special work the interests of this work and of the public are too often made subordinate to the interests of the naval or military service, more especially in the matter of change of station. For example, civil engineers object to the policy of placing river and harbor improvements in the hands of Army engineers, because one of the objects kept in view by the War Department in making details for this purpose is to vary the duty of the individual officer from time to time so as to give him a wider experience. Hence it may happen that an officer placed on duty in connection with the improvement of certain harbors on the Great Lakes shall, after three or four years, and just as he has gained sufficient experience of the peculiarities of lake work to make his supervision there peculiarly valuable, be transferred to work on the improvement of the Lower Mississippi with which he may be quite unfamiliar.

In like manner Professor Clarke objects to having a laboratory connected with the medical department of the Navy on the ground that the officer in charge is changed every three years; consequently science suffers in order that naval routine may be preserved.

There is force in this class of objections, but the moral I should draw from them is, not that Army and Navy officers should not be allowed to do work outside their own departments or in science, but that when they are put upon such duty, the ordinary routine of change of station every three or four years should not be enforced upon them without careful consideration of the circumstances of the case, and satisfactory evidence that the work on which they are engaged will not suffer by the change. And, as a matter of fact, I believe this has been the policy pursued, and instances could be given where an officer has been kept twenty years at one station for this very reason.

I pass over a number of objections that I have heard made to the employment of Army and Navy officers as administrators, on the ground that they are too "bumptious," or "domineering," or "supercilious," or "finicky," because every one knows what these mean and their force. An Army officer is not necessarily a polished gentleman; neither is a civilian; and a good organizer and administrator, whether officer or civilian, will at times, and especially to some people, appear arbitrary and dictatorial.

There is another objection to special details of Army or Navy officers for scientific duties which comes not so much from outside persons as from the War Department and the officers themselves, and it is this: Among such officers there are always a certain number who not only prefer special details to routine duty, but who actively seek for such details, who are perpetual candidates for them.

The proportion of men whose ideas as to their own scientific acquirements, merits, and claims to attention are excessive as compared with the ideas of their acquaintances on the same points is not greater in the Army than elsewhere, but when an Army officer is afflicted in this way the attack is sometimes very severe, and the so-called influence which he brings to bear may cause a good deal of annoyance to the Department, even if it be not sufficient to obtain his ends. I have heard officers of high rank, in a fit of impatience under such circumstances, express a most hearty and emphatic wish that no special details were possible, so that lobbying for them should be useless. This, however, seems to me to be too heroic a remedy for the disease, which, after all, only produces comparatively trifling irritation and discomfort.

The same evil exists, to a much greater extent, in the civil branches of the Government. Few persons can fully appreciate the loss of time, the worry, and the annoyance to which the responsible heads of some of our bureaus for scientific work are subjected through the desire of people for official position and for maintenance by the Government. They have to stand always at the bat and protect their wickets from the balls which are bowled at them in every direction, even from behind by some of their own subordinates.

It is true that a great majority of the balls go wide and cause little trouble, and a majority of the bowlers soon get tired and leave the field, but there are generally a few persistent ones who gradually acquire no small degree of skill in discovering the weak or unguarded points, and succeed in making things lively for a time. Considered from the point of view of the public interests, such men are useful, for although they cause some loss of valuable time, and occasionally do a little damage by promoting hostile legislation, yet their criticisms are often worth taking into account; they tend to prevent the machine from getting into a rut, and they promote activity and attention to business on the part of administrative chiefs. It is a saying among dog fanciers that a few fleas on a dog are good for him rather than otherwise, as they compel him to take some exercise under any circumstances.

At all events I think it very doubtful whether the jealousies and desire for position for one's self or one's friends which exist under present circumstances would be materially diminished under any other form of organization, even under a department of science.

Some conflict of interests now exists it is true; some work is duplicated ; but neither the conflict nor the duplication are necessarily wholly evil in themselves, nor in so far as they are evil are they necessary parts of the present system. This system is of the nature of a growth ; it is organic and not a mere pudding-stone aggregation of heterogeneous materials, and the wise course is to correct improper bendings and twistings gradually, prune judiciously, and go slow in trying to secure radical changes lest death or permanent deformity result.

It will be seen that in what I have said I have not attempted to eulogize science or scientists in the abstract. I should be very sorry, however, to have given any one the impression that I think they should not be eulogized. Having read a number of eloquent tributes to their importance by way of inducing a proper frame of mind in which to prepare this address, it is possible that I overdid it a little, and was in a sort of reaction stage when I began to write. But the more I have thought on the subject, and the more carefully I have sought to analyze the motives and character of those of my acquaintances who are either engaged in scientific work or who wish to be considered as so doing, and to compare them with those who have no pretensions to science, and who make none, the more I have been convinced that upon the whole the eulogium is the proper thing to give, and that it is not wise to be critical as to the true inwardness of all that we see or hear.

At least nine-tenths of the praises which have been heaped upon scientific men as a body are thoroughly well deserved. Among them are to be found a very large proportion of true gentlemen, larger, I think, than is to be found in any other class of men—men characterized by modesty, unselfishness, scrupulous honesty, and truthfulness, and by the full performance of their family and social duties.

Even their foibles may be likable. A little vanity or thirst for publicity, zeal in claiming priority of discovery, or undue wrath over the other scientist's theory, does not and should not detract from the esteem in which we hold them. A very good way of viewing characteristics which we do not like is to bear in mind that different parts of the brain have different functions; that all of them cannot act at once, and that their tendencies are sometimes contradictory.

There are times when a scientific man does not think scientifically, when he does not want to so think, and possibly when it is best that he should not so think. There is wisdom in Sam. Lawson's remark that "Folks that are always telling you what they don't believe are sort o' stringy and dry. There ain't no 'sorption got out o' not believing nothing." At one time the emotional, at another the intellectual, side of the scientific man has the ascendency, and one must appeal from one state to the other. Were scientific thinking rigorously carried out to practical results in every-day life there would be some very remarkable social changes, and perhaps some very disagreeable ones.

That scientific pursuits give great pleasure without reference to their utility, or to the fame or profit to be derived from them; that they tend to make a man good company to himself and to bring him into pleasant associations is certain; and that a man's own pleasure and happiness are things to be sought for in his work and companionship is also certain. If in this address I have ventured to hint that this may not be the only, nor even the most important, object in life, that one may be a scientific man, or even a man of science, and yet not be worthy of special reverence; because he may be at the same time an intensely selfish man, and even a vicious man, I hope that it is clearly understood that it is with no intention of depreciating the glory of science or the honor which is due to the large number of scientific gentlemen whom I see around me.

A scientific gentleman—all praise to him who merits this title— it is the blue ribbon of our day.

We live in a fortunate time and place; in the early manhood of a mighty nation, and in its capital city, which every year makes more beautiful, and richer in the treasures of science, literature, and art which all the keels of the sea and the iron roads of the land are bringing to it. Life implies death; growth presages decay; but we have good reasons for hoping that for our country and our people the evil days are yet far off. Yet we may not rest and eat lotus; we may not devote our lives to our own pleasure, even though it be pleasure derived from scientific investigation. No man lives for himself alone; the scientific man should do so least of all. There never was a time when the world had more need of him, and there never was a time when more care was needful lest his torch should prove a firebrand and destroy more than it illuminates.

The old creeds are quivering; shifting; changing like the colored flames on the surface of the Bessemer crucible. They are being analyzed, and accounted for, and toned down, and explained, until many are doubting whether there is any solid substratum beneath;

but the instinct which gave those creeds their influence is unchanged.

The religions and philosophies of the Orient seem to have little in common with modern science. The sage of the east did not try to climb the ladder of knowledge step by step. He sought a wisdom which he supposed far superior to all knowledge of earthly phenomena obtainable through the senses. The man of science of the west seeks knowledge by gradual accumulation, striving by comparison and experiment to eliminate the errors of individual observations, and doubting the possibility of attaining wisdom in any other way. The knowledge which he has, or seeks, is knowledge which may be acquired partly by individual effort and partly by co-operation, which requires material resources for its development, the search for which may be organized and pursued through the help of others, which is analogous in some respects to property which may be used for power or pleasure. The theologian and the poet claim that there is a wisdom which is not acquired but attained to, which cannot be communicated or received at pleasure, which comes in a way vaguely expressed by the words intuition or inspiration, which acts through and upon the emotional rather than the intellectual faculties, and which, thus acting, is sometimes of irresistible power in exciting and directing the actions of individuals and of communities.

The answer of the modern biologist to the old Hebrew question, *viz.*, "Why are children born with their hands clenched while men die with their hands wide open?" would not in the least resemble that given by the Rabbis, yet this last it is well that the scientist should also remember: "Because on entering the world men would grasp everything, but on leaving it all slips away." There exist in men certain mental phenomena, the study of which is included in what is known as ethics, and which are usually assumed to depend upon what is called moral law. Whether there is such a law and whether, if it exists, it can be logically deduced from observed facts in nature or is only known as a special revelation, are questions upon which scientific men in their present stage of development are not agreed. There is not yet any satisfactory scientific basis for what is recognized as sound ethics and morality throughout the civilized world; these rest upon another foundation.

This procession, bearing its lights of all kinds, smoky torches, clear-burning lamps, farthing rush-lights, and sputtering brimstone matches, passes through the few centuries of which we have a record, illuminating an area which varies, but which has been growing steadily larger.   The individual members of the procession come from, and pass into, shadow and darkness, but the light of the stream remains.   Yet it does not seem so much darkness, an infinite night, whence we come and whither we go, as a fog which at a little distance obscures or hides all things, but which, nevertheless, gives the impression that there is light beyond and above it.   In this fog we are living and groping, stumbling down blind alleys, only to find that there is no thoroughfare, getting lost and circling about on our own tracks as on a jumbie prairie; but slowly and irregularly we do seem to be getting on, and to be establishing some points in the survey of the continent of our own ignorance.

In some directions the man of science claims to lead the way; in others the artist, the poet, the devotee.   Far reaching as the speculations of the man of science may be, ranging from the constitution and nature of a universal protyle, through the building of a universe to its resolution again into primal matter or modes of motion, he can frame no hypothesis which shall explain consciousness, nor has he any data for a formula which shall tell what becomes of the individual when he disappears in the all-surrounding mist.   Does he go on seeking and learning in other ways or other worlds?   The great mass of mankind think that they have some information bearing on these questions; but, if so, it is a part of the wisdom of the Orient, and not of the physical or natural science of the Occident. Whether after death there shall come increase of knowledge, with increase of desires and of means of satisfying them, or whether there shall be freedom from all desire, and an end of coming and going, we do not know; nor is there any reason to suppose that it is a part of the plan of the universe that we should know.   We do know that the great majority of men think that there are such things as right and duty—God and a future life—and that to each man there comes the opportunity of doing something which he and others recognize to be his duty.   The scientific explanation of a part of the process by which this has been brought about, as by natural selection, heredity, education, progressive changes in this or that particular mass of brain matter, has not much bearing on the practical question of " What to do about it?"   But it does, nevertheless, indicate that it is not a characteristic to be denounced, or opposed, or neglected, since, even in the "struggle-for-existence" theory, it has been, and still is, of immense importance in human social development.

"Four men," says the Talmud, "entered Paradise. One beheld and died. One beheld and lost his senses. One destroyed the young plants. One only entered in peace and came out in peace." Many are the mystic and cabalistic interpretations which have been given of this saying; and if for "Paradise" we read the "world of knowledge" each of you can no doubt best interpret the parable for himself. Speaking to a body of scientific men, each of whom has, I hope, also certain unscientific beliefs, desires, hopes, and longings, I will only say: "Be strong and of a good courage." As scientific men, let us try to increase and diffuse knowledge; as men and citizens, let us try to be useful; and, in each capacity, let us do the work that comes to us honestly and thoroughly, and fear not the unknown future.

When we examine that wonderful series of wave marks which we call the spectrum we find, as we go downwards, that the vibrations become slower, the dark bands wider, until at last we reach a point where there seems to be no more movement; the blackness is continuous, the ray seems dead. Yet within this year Langley has found that a very long way lower down the pulsations again appear, and form, as it were, another spectrum; they never really ceased, but only changed in rhythm, requiring new apparatus or new senses to appreciate them. And it may well be that our human life is only a kind of lower spectrum, and that, beyond and above the broad black band which we call death, there are other modes of impulses—another spectrum—which registers the ceaseless beats of waves from the great central fountain of force, the heart of the universe, in modes of existence of which we can but dimly dream.

# BULLETIN

OF THE

# PHILOSOPHICAL SOCIETY OF WASHINGTON.

———

# GENERAL MEETING.

# BULLETIN

OF THE

# GENERAL MEETING.

---

**279TH MEETING.**                    **JANUARY 16, 1886.**

### President BILLINGS in the Chair.

Thirty-two members and guests present.

Announcement was made of the election to membership of Messrs. BENJAMIN PICKMAN MANN and CHARLES COOPER NOTT.

The following report of the Auditing Committee was presented by its chairman, Mr. TONER:

DECEMBER 24, 1885.

The undersigned, a committee appointed at the annual meeting of the Philosophical Society of Washington, December 19, 1885, for the purpose of auditing the accounts of the Treasurer, beg leave to report as follows:

We have examined the statement of receipts, including annual dues, sale of Bulletin, and interest on bonds, and find the same to be correct as stated.

We have examined the statement of disbursements, and compared the same with the vouchers, and find them to agree.

We have examined the returned checks and the bank account with Riggs & Co., and find the balance, $484.02, to agree with the statements in the Treasurer's report.

We have examined the U. S. bonds belonging to the Society, and find them to be in amount and character as represented in the Treasurer's report, aggregating $2,500.

> J. M. TONER,
> O. T. MASON,
> T. C. MENDENHALL,
> *Committee.*

(3)

Mr. J. S. DILLER communicated

## NOTES ON THE GEOLOGY OF NORTHERN CALIFORNIA.

[Abstract.]

Under the direction of Capt. Dutton I have spent the last three summers studying the geology of northern California and the adjacent portion of Oregon. The conclusions of a general nature referring to that region may be briefly summarized as follows:

In the northern end of the Sierra Nevada and the central portion of the Coast range, among the highly plicated, more or less metamorphosed strata which are older than those of the Chico group, there appears to be but one horizon of limestone, and that is of Carboniferous age.

The northern end of the Sierra Nevada is made up of three tilted orographic blocks which are separated from each other by great faults. The westernmost of these blocks stretching far to the southeast appears to form the greater portion of the range.

As in the Great Basin region the depressed side of each block was occupied by a body of water of considerable size. The deposits formed in these lakes gave rise to the fertile soils of American and Indian valleys.

The plication of the strata in the Sierra Nevada range took place, at least in great part, about the close of the Jurassic or beginning of the Cretaceous period, but the faulting which really gave birth to the Sierra as a separate and distinct range by differentiating it from the great platform stretching eastward into the Great Basin region, did not take place until towards the close of the Tertiary or the beginning of the Quaternary.

Although the faulting may have commenced earlier, the greater portion of the displacement has taken place since the deposition of a large part of the auriferous gravels and the beginning of the great volcanic outbursts in the vicinity of Lassen's Peak. If we may accept numerous small earthquake shocks as evidence, the faulting still continues.

The distribution of the rocks of the Chico group indicates that the western coast of the continent at that time lay along the western base of the Sierra extending around the northern end of the range in the vicinity of Lassen's Peak and stretching far northeasterly into Oregon. Off the coast lay a large island which now forms northwestern California and the adjacent portion of Oregon. This island extended as far southeast as the Pit river region where it was separated from the main land by a wide strait.

All of the ridges developed out of the Cretacean island belong to the Coast range.

The volcanic ridge of Lassen's Peak lies between the northern end of the Sierra Nevada and the Coast range. The great volcanic field of Oregon and Washington Territory, to which Lassen's Peak and the Cascade range belong, appears in a general way to be outlined by the depression between the Cretacean island and the main land. A general account of the facts from which these conclusions are drawn will appear in Bulletin of the U. S. Geological Survey No. 33.

Mr. I. C. RUSSELL read a supplementary paper entitled

NOTES ON THE FAULTS OF THE GREAT BASIN AND OF THE EASTERN BASE OF THE SIERRA NEVADA.

[Abstract.]

The structure of the Great Basin was systematically studied by the geologists of the Fortieth Parallel Exploration, and subsequently by G. K. Gilbert and J. W. Powell. The results of these investigations, so far as they relate to the faults of the region, are indicated in the bibliographic list which follows.

The studies here referred to led to the recognition of a type of mountain structure named the "Great Basin system," which has been found to prevail over large portions of the United States west of the Rocky Mountains. A typical mountain of this system is a long, narrow orographic block, upraised along one edge, i. e. a monoclinal ridge. A mountain range having this structure usually presents an abrupt scarp, formed of the edges of broken strata, on the side bordered by the fault, and slopes much more gently in the opposite direction.

Mountain ranges of this character occupy the greater part of the area of interior drainage, known as the Great Basin, and at times overlap its borders. An older structure in which corrugation plays an important part has been recognized by several geologists in the desert ranges of Nevada and Utah, but these disturbances were produced previous to the faulting which gave origin to the present topographic relief.

The writer has observed Great Basin structure to extend throughout Western Utah, Northern Nevada, and into Oregon as far as Malheur Lake. On the west side of the Great Basin, at the immediate base of the Sierra Nevada, there is an immense compound displacement that can be followed all the way from Honey Lake on the north to beyond Owen's Lake on the south, a distance of over 350 miles. Along many of the faults composing this belt the records of a post-Quaternary movement may be clearly recognized.. Fault scarps produced by recent movement have been observed in Eagle and Carson Valleys, south of Carson City, in Bridgeport Valley, and on the west side of Mono Lake. The earthquake in Owen's Valley in 1872, was caused by a movement along one of the faults of this series.

The eastern face of the Sierra Nevada is extremely abrupt and its western slope is gentle. Corrugations of older date than the faults which determine the present relief of the mountains may be observed at many localities. It thus agrees in its general features with many of the Basin ranges. The Sierra Nevada is essentially monoclinal in structure, but is traversed from north to south by faults which divide it into separate ranges, as may be seen in the neighborhood of Lake Tahoe and in the elevated region west of Mono Lake. The Great Basin structure here extends beyond the borders of the area of interior drainage, and is probably limited on the west by the great valley of California. How far north of Lake Tahoe the secondary faults that divide the mountain mass may be traced is unknown, but they can certainly be followed to where the Central Pacific railroad crosses the mountains.

The following list indicates where observations on the faults of the Great Basin system may be found:

Clarence King: Reports of the Fortieth Parallel Exploration. Vol. I, 1878, pp. 735, 744–746; Vol. III, 1870, p. 451.

J. D. Whitney: The Owens Valley earthquake. Overland Monthly, Aug. and Sept., 1872.

Joseph Le Conte: On the Structure and Origin of Mountains, with special reference to recent objections to the "Contraction Theory." American Journal of Science, Vol. XVI, 1878, pp. 95–112.

"          "          A theory of the formation of the great features of the earth's crust. American Journal of Science, Vol. IV, 1872, pp. 345–355, 460–472.

G. K. Gilbert: Progress report upon Geographical and Geological Explorations and Surveys West of the 100th Meridian, in 1872. Washington, 1874. p. 50.

"        "        Report upon Geographical and Geological Explorations and Surveys west of the 100th Meridian. Washington, 1875. Vol. III, Geology, pp. 21–42.

"        "        Contributions to the history of Lake Bonneville. In Second Annual Report of the U. S. Geological Survey. Washington, 1882. pp. 192–200.

"        "        A theory of the earthquakes of the Great Basin with a practical application. American Journal of Science, Vol. XXVII, 1884, pp. 49–53.

J. W. Powell: Basin Range System. See Report on Lands of the Arid Region of the United States. Washington, 1879. pp. 94–95.

"        "        Basin Range Province. See Report on the Geology of the eastern portion of the Uinta Mountains. Washington, 1876. pp. 6–7, 23–25.

C. E. Dutton: Geology of the High Plateaus of Utah. Washington, 1880. pp. 51–53.

I. C. Russell: Sketch of the Geological History of Lake Lahontan. Third Annual Report of the U. S. Geological Survey. Washington, 1883. p. 202.

"        "        A Geological Reconnoissance in Southern Oregon. Fourth Annual Report of the U. S. Geological Survey. Washington, 1884. pp. 442–453.

"        "        Lake Lahontan. Monograph No. XI, U. S. Geological Survey, pp. 24–28, 274–284.

Mr. GILBERT remarked that the section exhibited by Mr. Diller appeared to demonstrate a history comprising (1) the folding of the slates and the formation of several faults and associated monoclines, (2) the general degradation of the country until the monoclinal ridges were approximately obliterated, and (3) a renewal of movement on the old fault lines, giving rise to the existing topography.

Mr. WILLIS remarked that in 1883 he had had opportunity to study the Cascade Mountains north of the region described by Mr. Diller. The Sierra structure is apparently not found in the northern part of Washington Territory, and the eastern face of the Cascade range is probably not characterized like the Sierra by a great fault.

Mr. DILLER concurred in the statement that the Cascade range is built essentially of igneous rocks, and is not characterized by great faults, at least along its eastern base.

The topography of the Sierra has entirely changed since the deposition of the auriferous gravels, and some of the fault movements are so recent that the stream terraces to which they have given rise are still preserved.

Mr. G. K. GILBERT made a communication on

RECENT CHANGES OF LEVEL IN THE BASIN OF LAKE ONTARIO.

[The substance of this communication was presented to the American Association for the Advancement of Science at Ann Arbor, and appears in abstract in Science, Vol. VI, p. 222.]

Remarks were made by Mr. E. FARQUHAR.

---

280TH MEETING.                    JANUARY 30, 1886

The President in the Chair.

Fifty-five members and guests present.

The Chair announced the appointment of the Committee on Communications.

Mr. GEORGE E. CURTIS made a communication on

LIEUTENANT LOCKWOOD'S EXPEDITION TO FARTHEST NORTH.

[Abstract.]

The paper opened with a reference to the statement in the Encyclopædia Britannica (article, Polar Regions, p. 326,) that "all this region [the northern coast of Greenland and the interior of Grinnell

Land] had already been explored and exhaustively examined by the English expedition of 1875–'76." A refutation of this statement was not now necessary inasmuch as a retraction had already been made; but an impartial examination of Lieut. Lockwood's observations was still required as a basis for our own confidence in the latitude attained.

A description of the equipment of the expedition was given, with a sketch of the events of the journey, and extracts from the narrative report. The weights of the food and equipments drawn by the dog team furnished the basis of a discussion as to the value of dogs in arctic sledging. The weight of food taken for the support of each man was about twice that taken for each dog. Now if a man can drag a sufficiently greater amount to compensate for the greater weight of his food, it is immaterial whether the motive force used be dogs or men. On this expedition the dog sledge was actually loaded so as to give a weight of about 100 lbs. to each dog; but the maximum weight that can be advantageously drawn by a man is only 125 or, perhaps, 150 lbs. The ratio of effective work performed to the weight of food consumed is, therefore, materially greater for dogs than for men, so that a substantial economical advantage is obtained by using dogs instead of men for sledge dragging. This advantage seems not to have been appreciated by the English expedition of 1875'–76, whose heavy sledges and equipments were all drawn by hand. In addition to the more conspicuous causes of the failure of Lieut. Beaumont's expedition on the Greenland coast, the neglect to make use of dogs must be added as an important element.

The sextant observations made by Lieut. Lockwood for determining the position of his farthest north were shown to be highly satisfactory. Sets of circum-meridian observations for latitude were made at midnight of May 14th and at noon of May 15th. The conditions of observation offer no sufficient reason for giving more weight to one set than to the other. The mean of these results gives 83° 24' as the latitude attained by Lieut. Lockwood, and an uncertainty not greater than 1' represent the accuracy of its determination.

The paper closed with the following tribute to the character of Lieut. Lockwood as an arctic explorer:

I cannot close this review of Lieut. Lockwood's expedition to farthest north without turning from the cold discussion of the astronomical and geographical records to speak of him of whose life and labors they constitute an imperishable memorial.

The success of the expedition was not the result of chance, but was due to Lieut. Lockwood's thorough knowledge of the details of such an undertaking, and to his indomitable energy in its execution. During the preceding winter he had devoted himself to preparation for the work; had made a careful study of the management and equipment of previous sledging expeditions—especially those of the English in 1875–'76—and profiting by the experience of his predecessors was able to avoid their mistakes. Lieut. Beaumont's journey on the Greenland coast was impeded by the heavy sledge, and the heavier equipments with which it was weighted. Lieut. Lockwood's extraordinary distance was attained with a light sledge drawn by dogs and loaded with nothing but food and the barest necessities of a camp. Regardless of all personal comforts, everything was sacrificed to the objects of the expedition.

Under the instruction of Mr. Israel, the young astronomer, Lieut. Lockwood had familiarized himself during the winter with all the astronomical observations necessary to be made by an explorer, and with the return of the spring sun applied himself to practical observations with the sextant until he became an expert in its use. So good was his astronomical work that the accuracy of his observations is dependent only on the variability of the instrument and the difficulty of the conditions of observation.

In addition to a practical knowledge of arctic sledging, the expedition was undertaken with a determined energy of purpose, those qualities expressively termed "grit" and "pluck," which no obstacle could defeat. Retaining only two companions at Cape Bryant, he sent back his supporting party and continued his advance over an unknown coast. Suffering continuously from cold, hunger, or fatigue, he pushed on with unflinching perseverance until one hundred and fifty miles of new coast were traversed and the national colors unfurled in the highest latitude ever attained by man.

Simply to go a little nearer the pole than his predecessors was not, however, the controlling object of this expedition. Lieut. Lockwood's own motives, as we read them in his journal, were these: "My great wish is to accomplish something on the north coast of Greenland that will reflect credit on myself and on the expedition." Inspired by this praiseworthy ambition, his skillful management resulted in its most successful realization. His motives were not those of the visionary and enthusiast who "knows nothing and fears nothing," but of an earnest practical explorer whose ambition is to add something to the world's knowledge of the planet on which we live. The literal fidelity of his narrative, its freedom from an exaggeration that has too often marred the records of previous Arctic explorers, the exact and painstaking descriptions, and the careful distinction between what is seen and what is inferred, all bear witness to his conscientiousness in the search for truth.

As an important element in the success of Lieut. Lockwood's expeditions, due recognition must be given to the cordial, sympathetic and able co-operation of Sergeant Brainard. Chosen by Lieut. Lockwood to continue the journey to Cape Bryant, when all the remainder of the party returned, it was Brainard who pushed onward with him over one hundred and fifty miles of that desolate coast and reached the farthest north. It was likewise Lockwood and Brainard who a year later, in May, 1883, explored the interior of Grinnell Land and looked out on the shores of the western polar sea.

But only one of these companions in exploration was destined to reach home to receive the honor due to their heroic achievements—honor *due*, but, as yet, awarded neither to the living nor the dead. The story of the return is known to all, but perhaps not Lieut. Lockwood's wonderful cheerfulness of spirit through that last terrible winter at Cape Sabine with death staring him in the face Lieut. Lockwood died on April 9, 1884, "from action of water on the heart induced by insufficient nutrition"—the official euphemism for starvation. This record of indescribable suffering, privation and death, following that of two years of heroic endeavor and achievement, is a tragedy which appeals to human hearts with a force unequalled by any story of fiction or by any drama of the stage.

To Lieut. Lockwood's achievements are applicable the familiar lines of Horace :

> " Exegi monumentum aere perennius
> Regalique situ pyramidum altius,
> Quod non imber edax, non Aquilo impotens
> Possit diruere, aut innumerabilis
> Annorum series et fuga temporum."

Woven into the history of arctic discovery and engraven on our maps, the substantial results of Lieut. Lockwood's explorations form a tablet more enduring than brass, which the corroding storm, the fierce north wind, and the flight of ages cannot efface.

In reply to a question by Mr. Mussey, Mr. CURTIS stated that the time for longitude determination was obtained from one ordinary watch of good quality, and one pocket chronometer. Messrs. DALL and ROBINSON discussed the advantages and disadvantages of the use of dogs in arctic sledging, and attention was called to the importance of using snow shoes, and of coating the sledge runners with ice.

Mr. O. T. MASON made a communication on

TWO EXAMPLES OF SIMILAR INVENTIONS IN AREAS WIDELY APART.

[Abstract.]

Anthropologists assign similar inventions observed in different parts of the world to one of the following causes:

1. The migration of a certain race or people who made the invention. Upon this theory similar inventions argue the presence of the same people or race.

2. The migration of ideas—that is, an invention may be made by a certain race or people and taught or loaned to peoples far removed in time and place. Upon this theory similar inventions argue identity of origin, but not necessarily the consanguinity of those who practice them.

3. In human culture, as in nature elsewhere, like causes produce like effects. Under the same stress and resources the same inventions will arise.

Now, the question arises, which of these causes shall be invoked in specific cases to account for resemblances.

We must first examine the word resemblance.

Taking Aristotle's four causes:

The material cause, *ex qua aliquid fit.*

The formal cause, *per quam.*

The efficient cause, *a qua.*

The final cause, *propter quam.*

We must enlarge upon them as follows: Every human activity involves six fundamental considerations.

1. The agent, or efficient cause.

2. The material cause.

3. The implemental cause.

4. The formal cause.

5. The processive cause—that is, the exact order and method of the action.

6. The motive or function.

We might, also, include a series of concomitants, such as technical vocabulary, all sorts of traditional lore and myths, social organization, and even religious rites.

Again, some of the six causes are themselves generally the outcome of other causes, so that we have concatenations and genealogies of causes.

Now for the application. Most men, when they say this thing resembles that, have reference only to one of our six causes. They mean simply that there is resemblance in form, or material, or technical method, or function. My plan would be to submit such resemblances to scrutiny to ascertain how far they extend, and, also, to examine resemblances known to be consanguine, or borrowed, or independent, to ascertain which of our characteristics are peculiar to them. In that way an inductive system of rules would be adduced.

The two independent inventions which I exhibit are a beginning in that direction. One is a stitch in basketry, found only at Cape Flattery and on the Congo. This stitch is common enough in fish-traps, wattling fences, and cages, but in only these two areas have people thought to apply it to close basketry. It consists of vertical warp, a horizontal second warp, laid behind the first, and a coiling or sewing of these two together, so as to show a diagonal stitch in front and a vertical stitch in the rear. Here the resemblance is in method alone. In all other respects the inventions differ.

The other invention referred to is the throwing-stick of Australians, Puru Purus, and Eskimo. These agree, in motive or function and in the fundamental idea of a staff and a' hook. Beyond this the Eskimo have invented a dozen additional attachments never dreamed of by the others.

Mr. Murdoch supplemented the enumeration of throwing sticks by describing an undeveloped form used by the Siberian Eskimo. In reply to a question by Mr. Goode, Mr. Mason stated that he had not seen the Brazilian sticks; they are mentioned by many travelers.   Mr. MANN and Mr. MURDOCH described the manner in which the throwing stick is used by Eskimo in kyaks.   The motion centers in the wrist and not the elbow or shoulder.

---

281ST MEETING.                                FEBRUARY 13, 1886.

### The President in the Chair.

Fifty-five members and guests present.

Mr. J. H. KIDDER communicated an

HISTORICAL SKETCH OF DEEP SEA TEMPERATURE OBSERVATIONS,

illustrating the subject by numerous diagrams and by a collection of deep sea thermometers.

Mr. E. B. ELLIOTT made a communication on the

ANNUAL PROFIT TO BANKS OF NATIONAL BANK NOTE CIRCULATION,

and a second communication on the

QUANTITY OF UNITED STATES SUBSIDIARY SILVER COIN EXISTING AND IN CIRCULATION.

In these papers he developed the formulæ used in computing certain tables embodied in the report of the Comptroller of the Currency.

Remarks were made by Messrs. MUSSEY and LAWRENCE.

Mr. ASAPH HALL read a paper on

THE NEW STAR IN THE NEBULA OF ANDROMEDA,

giving an historical account of its discovery, growth and decadence. [This paper is printed in the American Journal of Science, 3d series, vol. XXXI, p. 299.]

282D MEETING.                    FEBRUARY 27, 1886.

The President in the Chair.

Fifty-six members and guests present.

The Chair announced the election to membership of Mr. GEORGE JOTHAM CUMMINGS.

Mr. ASAPH HALL made a communication on

THE IMAGES OF STARS,

which was discussed by Messrs. EASTMAN, CURTIS, and PAUL. [This paper is published in the Sidereal Messenger, April, 1886.]

Mr. R. S. WOODWARD made a communication

ON THE CHANGES OF TERRESTRIAL LEVEL SURFACES DUE TO VARIATIONS IN DISTRIBUTION OF SUPERFICIAL MATTER.

[To appear as a Bulletin of the U. S. Geological Survey.]

He was followed by Mr. G. K. GILBERT with a paper

ON THE OBSERVED CHANGES OF LEVEL SURFACES IN THE BONNE-VILLE AREA, AND THEIR EXPLANATION;

and Mr. T. C. CHAMBERLIN then began a paper

ON THE VARYING ATTITUDES OF FORMER LEVEL SURFACES IN THE GREAT LAKE REGION AND THE APPLICABILITY OF PRO-POSED EXPLANATIONS.

---

283D MEETING.                    MARCH 13, 1886.

Vice-President HARKNESS in the Chair.

Thirty-nine members and guests present.

The Secretary read a letter from the Secretary of the Council of the Anthropological Society, inviting the members of the Philo-sophical Society and their friends to attend the annual meeting of the Anthropological Society and listen to an address by its president, Major J. W. POWELL.

The Chair announced the election to membership of Messrs. CARLOS ALBERT KENASTON, ROLAND DUER IRVING and ARTEMAS MARTIN.

Mr. T. C. CHAMBERLIN completed his communication

## ON THE VARYING ATTITUDES OF FORMER LEVEL SURFACES IN THE GREAT LAKE REGION AND THE APPLICABILITY OF PROPOSED EXPLANATIONS.

Remarks were made by Mr. DUTTON.

Mr. R. D. IRVING made a communication on

## THE ENLARGEMENT OF MINERAL FRAGMENTS AS A FACTOR IN ROCK ALTERATION,

which was discussed by Messrs. IDDINGS, DILLER, DUTTON, and LAWRENCE.

---

284TH MEETING.                                    MARCH 27, 1886.

### The President in the Chair.

Thirty-three members and guests present.

Mr. I. C. RUSSELL made a communication on

## THE SUBAERIAL DECAY OF ROCKS AND THE ORIGIN OF THE RED CLAY OF CERTAIN FORMATIONS.

This was discussed by Prof. JOHN S. NEWBERRY, of New York city, and by Messrs. GOODE, DARTON, IRVING, and CHAMBERLIN.

Mr. ROMYN HITCHCOCK made a communication on

## RECENT IMPROVEMENTS IN MICROSCOPIC OBJECTIVES, WITH DEMONSTRATION OF THE RESOLVING POWER OF A NEW 1-16TH INCH.

Remarks were made by the President.

Mr. HENRY FARQUHAR read a communication on

## A FONÉTIK ÆLFABET.

[Ǽbstrækt.]

△is ælfabet œndœrtéyks tu reprizént ðe sawndz av Íɣglic spiytc æz yúwjuwali hœrd, bai twénti-nain létœrz. Av ðiyz *b, d, f, g, h, k, l, m, n, p, r, s, t, v, w, y* ænd *z* hæv ðer kœstomeri sawndz; *c* hæz its sawnd æz in *benifícieri* [beneficiary] — a sawnd akéyjœnali gíven tu *s* or *t* or *ch,* or mœœr ófen tu *sh; j* iz kanfáind tu its Frentc sawnd — hwitc iz ool ðæt iz left tu it hwen kambáind wið *d* in Íɣglic, æz in *œ́djutœnt* [adjutant]; hwail ðe néyzal yúwjuwali ríten *ng,* ðe *th* flæt ænd *th* carp, ar gíven bai ðe létœrz ɣ, ð ænd ꝺ, bárod fram ðe Griyk ælfabet. Av ðe váwels, *a* iz æz in *wad* or *bar, e* æz in *pet, i* æz in *pit, o* æz in *on* or *or, u* æz in *put;* hwail ðe dáigræfs œ ænd œ ar yuwzd for ðe váwel sawndz hœrd in *bœt* [bat] ænd in *bœt* [but; won, burr, stir, herd, heard, word, etc.]. Dœblld létœrz índikeyt prolóoɣd sawndz: *l* in *dœ́bll* [double], *m* in *prízmm* [prism], *a* in *stáari* [starry], *o* in *doon* [dawn]. Œ́ðœr loɣ váwels ar rigárded æz impyúœr sawndz; ænd slœrz bifóœr *r* ar dinówted bai œ, téndensi tu klowz wið ðe lips bai *w,* ænd wið ðe tœɣ bai *y,* fúlowiɣ a cort váwel; dœs wiy hæv œœ, iœ, oœ, uœ in *beœr, biœr, boœr, buœr* [bear, beer, bore, boor], *aw, ow, uw* in *haws, flow, buwt* [house, flow, boot], *ey, iy, ay* in *beyt, biyt, bayt* [bait, beat, bite]. Æz egzæmpllz av œ́ðœr kàm-binéyœœnz, *oil, wain, yuwz, tényur, kyuœr* [oil, wine, use, tenure, cure] mey biy gíven. Æksénted váwelz ar markt, sékanderi æksents (æz on ðe fœrst sílabll av *vèriabíliti*) hæviɣ ðe greyv sain, príusipal wœnz ðe ækyúwt.

Ówiɣ tu ðe páwœrful ínfluwens av ðe skuwlz in kíypiɣ œp æn ek-sésiv révœrens for ðe kanvéncœnal stændard av spéliɣ, ænd tu ðe fækt ðæt ðe mædjáriti av ðe wœrdz av áwœr læɣgwedj ar nown tu œs bai ðer æpíœrens æz prínted or ríten rædœr ðæn bai ðer sawnd æz spówken, ðer iz lítll imíydiet prúspekt av sœksés in ðe "spéliɣ-rifórm" múwvment, hwitc eymz tu divélap áwœr spéliɣ intu wœn mœœr nícerli fonétik. Nèvœrðelés, ðer ar præktikal yúwsez for a

2

gud fonétik ælfabet.  Fœrst, it kud fœrnic a stændard av pronœnsiéy-
cœn in díkcœneriz av ool sorts — huwz ríydœrz wud ðœs biy seyvd
ðe tæsk av lœrniɤ a dífœrent sístem for iytc díkcœneri, æz æt préz-
ent.  Sékand, it kud ænd cud form a régyùlar bræntc av instrœk-
cœn in skuwlz, ænd dœs æsíst in sikyúœriɤ yúwniform pronœn-
siéycœn.  θœrd, sœm sœtc diváys iz olmowst a nisésiti, if wiy wud
briɤ órdœr íntu a bíznes naw òltugéðœr keyátik ; trænslitœréycœn av
neymz fram ðe Rúwcan, Ærabik ænd œðœr læɤgwedjez nat yúwziɤ
ðe Rówman ælfabet.  Æz wiy fálo now méðad kansístentli æt préz-
ent, ðer wud pœrhæps biy now greyt dífikœlti in estæbliciɤ wœn;
espécali if bai it ðe sayt av a neym kud giv æn æpráksimet aidía
haw it iz pronáwnst bai ðowz mowst fæmíliar wið it.

Mr. Mussey said that the principal difficulty in all phonetic
alphabets was not in the alphabets themselves, but in the existing
variety of pronunciation and the disagreements as to the true
pronunciation of words.  Pitman's phonetic system—little short
of an inspiration—was to his mind the best ever devised for
practical use, though Bell's system of visible speech enabled a
person familiar with it to correctly pronounce words and sentences
in any language whatever.

Mr. Mallery said that he had been connected with the preparation of a phonetic alphabet by the Bureau of Ethnology, with
the object of collecting and recording the vocabularies of the
languages of the North American Indians.  In addition to the
requisite that there should be a distinct character for every sound,
it was made a fundamental rule that the characters should be
limited to those in an ordinary font of English type, embracing
however not only the Roman alphabet but such characters and dia-
critical marks as the printers' cases of average newspapers could
furnish.  This was accomplished so as to provide for many more
sounds than are included in Mr. Farquhar's scheme, yet without
resort to the Greek letters used by him in several instances.  It
was done by the simple device of reversing the large number of
letters in the Roman alphabet which present a markedly different
appearance when so reversed, from their erect position.  This is
entirely convenient to the printer and does not occasion awkward-
ness in the current script to the recorder or writer for the press, as

it is only necessary to mark the letter intended to be reversed, after writing it in the normal manner, and to notify the printer accordingly. In practice the letters intended to be reversed are marked by a cross beneath them, though a still more current method of distinction would be by the cedilla in using which the pen or pencil is not removed from the letter as formed. This is however more convenient to the writer than to the printer.

The result of this scheme in practice has solved one part of the problem of a universal phonetic alphabet. Vocabularies and chrestomathies of unwritten languages have been recorded and printed, upon which grammars and dictionaries have also been prepared and printed, and from them the languages can be learned so as to be spoken intelligibly without oral instruction. The possibility of the use of such an alphabet with only such modification as would increase its simplicity, in the reform of the English literation, is not to be doubted, in view of its success under more difficult conditions. The actual obstacles to phonetic reform of fixed alphabets are, though perhaps insuperable, non-essential in the scientific view of the question.

Concerning the gliding sounds mentioned by Mr. Farquhar, Mr. DOOLITTLE remarked that some of these appeared to be essential, while others were only accidental.

---

285TH MEETING. APRIL 10, 1886.

The President in the Chair.

Fifty-nine members and guests present.

The Chair announced the election to membership of Messrs. ALEXANDER GEORGE MCADIE and ROBERT THOMAS HILL.

Mr. MALLERY read a communication on

CUSTOMS OF EVERY-DAY LIFE.

[Abstract.]

The scope of the paper excluded the more commonly noted ceremonial institutions, such as appear in regal courts, courts of justice, and legislative bodies, and embraced the ordinary modes of behavior among civilized people. These all have history and significance, are not the result of deliberate invention or convention, and in their present shape clearly exhibit the laws of evolution, though not always in the directions set forth in text-books and treatises on sociology. Comment was made upon the topics of social etiquette, precedence, titles, grammatical forms of personalty, the address and signatures of letters, forms and practices relating to written invitations and to social visits, and formularies of oral greeting, with examples or illustrations under each topic.

Fashion was distinguished from custom as being imitative and transitory, although in some few instances genuine merit in a fashion led to its permanent adoption under the same law with which the convenient and useful portions of old customs have survived in modifications.

Two points on which the paper specially declared disagreement with Herbert Spencer relate to the bow in salutation and to the hand-shake or grasp. The bow Mr. Spencer regards as but modified from the natural expressions of physical fear and bodily subjection noticed among sub-human animals and the lowest tribes of men, originating in actual prostration and groveling to which crawling and kneeling succeed, and the bow is but a simulated and partial prostration. A large class of obeisances doubtless had their origin in the attitudes of fear, and several were adduced in addition to those mentioned by Mr. Spencer, but it was contended that the subject of the bow is much more complex than as presented by him, a separate and independent course of evolution being suggested. Evidence was collected from many sources, and especially from gesture speech, relating to the concepts of, and expressions for, higher and lower, superior and inferior, assent, submission and respect, all connected with the forward and downward inclination of the head in salutation. Regarding the uncovering of the head as a part of the masculine bow, the paper offered to Mr. Spencer a new illustration of militancy, too often insisted upon in his Synthetic Philosophy but not definitely in this connection. The voluntary deprivation of removable head gear—once defensive—is often a mark of defeat and subjection. The modern formal military and naval salutes contain the same idea that the saluter is actually or symbolically powerless. Therefore the action of the removal of the hat, the present repre-sentative of the casque, helmet, or morion, is better adapted to a "surrender" theory than to that of pretended "beggary" advocated by Mr. Spencer.

That great writer believes that the hand-shake originated in a struggle, first real, afterwards fictitious, in which each of the performers attempted to kiss the hand of the other, which was resisted, thus producing a reciprocating movement of the joined hands. In examining this explanation the antiquity and prevalence of the kiss in salutation was questioned. The mutual kiss of affection or passion by the lips between opposite sexes is not found among the lower tribes and is probably not of great antiquity. It was preceded without reference to sex by patting, stroking or rubbing different parts of the body—smelling and sniffling being also common. The kiss of the hand is undoubtedly ancient and gestural, and is apparently not derived from that of the lips, which is gustatory. Instances were admitted of the identical friendly contest for priority in kissing hands relied on by Mr. Spencer, but they were considered to be connected with the topic of precedence as secondary, the joining of hands being primary and wholly unconnected with a "shake" or any motion after junction. Evidence was presented that the junction of hands in testimony and in expression of agreement and friendship is of too high antiquity and universality to be derived from a pantomimic contest about precedence for the comparatively modern and limited kiss.

Mr. Mason expressed the opinion that not all customs are indirectly derived, and cited the innovations of the day as instances of customs deliberately assumed for a definite purpose.

Mr. Mendenhall referred to the numerous ways in which Japanese customs are the inverse of ours. In beckoning, the fingers are turned down instead of up, and this is probably explained by the fact that in Japan those who are beckoned—namely, inferiors—are by custom or in theory prostrate. The kitchen of a hotel is placed at the front against the street. A horse is backed into the stable and led out. It is a matter of etiquette, and modesty also, that ladies turn their toes in. The Japanese do not shake hands; the bow is very low, and is begun twenty feet away. There is no kiss of ceremony or friendship, but the kiss exists as an expression of passion. Japanese children play all the common games of our children in some modified form, except marbles.

In reply to a question by Mr. Mason, Mr. Mendenhall said that a Japanese does not shake his own hands as a salutation to another, but may join them in bowing as a merely accidental attitude. Messrs. E. FARQUHAR and MUSSEY spoke of the antiquity of kissing as indicated by Hebrew and Greek literature. Other remarks were made by the President and by Mr. CLARKE.

Mr. R. D. MUSSEY made a communication entitled

WHEN I FIRST SAW THE CHOLERA BACILLUS.

---

286TH MEETING.                                    APRIL 24, 1886.

The President in the Chair.

Thirty members present.

The President communicated an invitation from the American Historical Association to attend its sessions of April 27–29.

Mr. G. BROWN GOODE and Mr. T. H. BEAN made a joint communication on

THE DISTRIBUTION OF FISHES IN THE OCEANIC ABYSSES AND MIDDLE STRATA.

Remarks were made by Messrs. PAUL, HARKNESS, BILLINGS, DOOLITTLE, GOODE, WELLING, and TAYLOR, and by Prof. EDWARD D. COPE, of Philadelphia.

Mr. GILBERT THOMPSON made a communication on

THE PHYSICAL-GEOGRAPHICAL DIVISIONS OF THE SOUTHEASTERN PORTION OF THE UNITED STATES AND THEIR CORRESPONDING TOPOGRAPHICAL TYPES.

[Abstract.]

Having charge of the geographical work carried on by the U. S. Geological Survey in that portion of the Appalachian region south of Pennsylvania and the Ohio river, I have had occasion to consider the classification of the region from the point of view of the geographer. It has previously been divided by many authors and into numerous sections, the basis of classification being geological botanical, agricultural, or commercial, and usually from a local standpoint. For my purposes the principal basis of classification is the character of the topographic relief, but this is so closely related to the features controlling other classifications that a large share of the boundaries coincide with lines previously drawn and the selection of appropriate designations is little more than a choice between names previously given.

There are some parts of the United States where the drainage basin affords the best unit for the purposes of the physical geographer. This holds for the basin of the Laurentian lakes, the basin of the Red River of the North, and the great Interior basin. But in the Appalachian region the drainage cannot be used. There is however in this region a remarkable line of demarcation, known as the *fall line*, which finds its manifestation in connection with the drainage, and is the natural boundary of an important division. If we follow the course of any river in the eastern part of the United States, south of New England, from its source to the sea, we discover that at a certain point it ceases to be rapid and turbulent, and becomes broad and slow-moving, and in many cases an estuary of the sea. At the point where this change occurs there is usually a fall or rapid. The familiar local example is the Potomac at Little Falls. I have traced this fall line from near Troy, N. Y., southward by the interior cities of Washington, Richmond, Columbia, and Montgomery, and thence to the Muscle Shoals of the Tennessee river. It is always the lower limit of water power and often the upper limit of navigation, and is therefore marked, and destined to be marked, by cities and towns of importance. In its northern portion it is at the head of tide, and nowhere does it exceed an altitude of 200 feet. It may yet be determined that it crosses the Saint Lawrence at the Lachine rapids and the Mississippi above Cairo, although no rapid exists at that point. Whether it may be traced farther and into Mexico remains to be determined.

From the fall line to the shore of the sea there is a region having a gentle slope, traversed by slow-moving rivers, and fringed at almost a dead level by deltas, swamps and everglades. This I have entitled the *coastal plains*, including as subdivisions the *Atlantic plains* and the *Gulf plains*.

The area bounded by the fall line and by the Mississippi and Ohio rivers and a part of the drainage divide of the Laurentian lakes, might be taken as a whole as the Appalachian region, but it includes three sections so distinct in topographic type as to warrant separate designations.

From the Ohio river southeastward and from the Mississippi

eastward the country gradually rises until it reaches an altitude of about 2500 feet above sea level, where it is generally cut off by an escarpment facing to the southeast and about 1200 feet in height. The composite name of Cumberland-Allegheny-Catskill plateau would serve to define it, but for brevity I have designated the whole as the *Cumberland plateau.* Its general topographic character is that of a table land deeply cut by a system of ramifying drainage. At the north the surface is somewhat rolling, and the plateau ends at the south in long, finger-like spurs. Its rivers and streams rise generally near the edge of the escarpment and flow toward the northwest. The Potomac however breaks across the edge and flows eastward, while the New and Tennessee rivers enter the plateau from the east and flow westward.

From the Cumberland plateau eastward to the eastern foot of the Blue Ridge lies a belt to which the name of *Appalachian region* is applied in a restricted and definite sense. It is characterized by numerous long, narrow mountain ridges, closely parallel to each other and bending in sympathy with the local curvature of the belt. Through large areas they are approximately uniform in height, but elsewhere they are unequal. In a notable belt, everywhere recognized in the local nomenclature as a valley, and traversing the region from north to south, the ridges are so low that they rank only as hills. At the north the principal mountain area lies west of the great valley and only the Blue Ridge on the east. At the south the valley lies close to the Cumberland plateau, and the Blue Ridge is expanded into a broad mountain district, culminating in Mt. Mitchell (6711 feet), the highest summit east of the Rocky Mountains.

The remaining area is the *Piedmont region,* an undulating plain, diversified by low spurs from the mountain region, and occasional isolated hills of considerable elevation. The streams are rapid, and the topographic relief gradually diminishes toward the fall line.

The communication was fully illustrated by maps and topographic sketches, and by a profile from Louisville, Kentucky, to Charleston, South Carolina. Remarks were made by Messrs. GILBERT, HARKNESS and COPE.

287TH MEETING.                                    MAY 8, 1886.

### The President in the Chair.

Thirty-five members present.

The Chair announced the election to membership of Messrs. JOSEPH HAMMOND BRYAN and MERWIN MARIE SNELL.

Mr. THOMAS RUSSELL made the following communication on

## TEMPERATURES AT WHICH DIFFERENCES BETWEEN MERCURIAL AND AIR THERMOMETERS ARE GREATEST.

Glass and mercury do not expand uniformly. An increase in temperature of one degree at one hundred degrees causes greater changes of volume than the same increase at zero. (In all references here to degrees and temperatures the centigrade scale is to be understood.) Normal mercurial thermometers, when corrected for their various errors of construction, differ among themselves and also from the air-thermometer.

At 40° the mercurial thermometer reads about 0.°2 higher than the air-thermometer. At — 38.°8, the melting point of mercury, it reads about 0.°2 lower. The quality of mercury in a thermometer has an influence on its reading. A thermometer containing $\frac{1}{10000}$ of lead in the mercury will read 0.°5 lower at 50° than if the mercury is pure. [H. J. Green.]

Comparisons have been made at the Signal Office between an air-thermometer and a number of mercurials. Some deductions have been made from the results of this work as to the temperatures at which the differences between the two thermometers are greatest. From the same results there have also been derived values of the coefficients of expansion of glass dependent on the second and third powers of the temperatures. It is to these I wish to call your attention.

The air-thermometer used was of the kind that measures temperatures by the varying pressure of a quantity of air kept at a constant volume. Five Tounelot mercurial thermometers were compared with this air-thermometer at temperatures from 0° to 55°. Two Baudin thermometers were compared with it from 0° to — 38.°8.

The freezing points of mercurial thermometers rise with age. A few days after a thermometer is filled this rise may amount to a whole degree.    In a year after that it may rise an additional five-tenths of a degree; with succeeding years the change is less and less.

When a thermometer is raised to a high temperature its freezing point is depressed.    The average depression for 100° is about 0.°2. On raising to a temperature lower than 100° the freezing point is also depressed, but not so much.    For 50° the depression is about 0.°05.    For temperatures as high as 100° the depressions are about proportional to the squares of the temperatures by which they are produced.

The cause of these changes of freezing point is in the nature of the glass.    The mercury in the thermometer has nothing to do with them; neither has the atmospheric pressure.

The amount of the changes depends on the composition of the glass in the thermometer-bulb.    It has been recently ascertained by H. F. Wiebe that the change is greatest for glass containing equal quantities of potash and soda.    A thermometer made of a variety of glass containing 14 per cent. of potash, 14 per cent. of soda, the remainder silica and oxide of lead, was found to have its freezing point depressed 0.°84 on raising it to 100°.    Thermometers in which the potash or soda in the glass was replaced by lime were found to have the freezing points depressed only 0.°07 for the temperature of 100°.

To produce the maximum depression of freezing point peculiar to any temperature requires that the thermometer be kept at that temperature for a certain length of time.    For a temperature of 100° a half hour suffices; for 50° two hours are required.

If the thermometer is kept at 100° longer than half an hour the depressed freezing point after that time begins to rise.    If continued at the higher temperature for two weeks the freezing point at the end of that time will be found to have risen about one degree. This fact is taken advantage of by some makers of thermometers to produce an instrument whose freezing point will vary but little in years subsequent to its manufacture.

The depression of freezing point produced by high temperature is only temporary.    The thermometer in the course of time regains the reading of its freezing point corresponding to ordinary temperatures.    The more quickly the depression is produced the more slowly the reading is regained.

After a thermometer has been subjected to a temperature of 100°

it will regain its ordinary freezing point reading in one month. The change in the first part of this period is much more rapid than towards the end. To recover the depression caused by 50° requires only two days. The older a thermometer the more quickly it gains its freezing point corresponding to ordinary temperatures. An instrument forty years old will regain its freezing point after exposure to 100° in one week while an instrument three years old requires a month.

The more alternations of temperature a thermometer is subjected to the more quickly its freezing point rises.

A thermometer subjected to a very high temperature, as 350°, will will have its freezing point raised from 12° to 20°. This rise is not due to softness of the glass at the high temperature and a consequent diminution in the volume of the bulb by the atmospheric pressure. This is shown by experiments with weight-thermometers. In these the tubes are open to the pressure of the air and there is as much pressure inside as outside the bulb.

As heating to 100° depresses the freezing point while heating to 350° raises it there must be some intermediate temperature for which there is no change. This point is usually at the temperature of about 160° to 180°, but varies widely with thermometers made of different kinds of glass.

When a thermometer is subjected to a very low temperature a temporary rise in its freezing point is produced. To produce an appreciable rise requires a long-continued exposure. After being kept twenty-four hours at — 30° the freezing point is found to be about 0.05° higher than at first.

One hundred degrees on the centigrade scale is taken as the temperature of steam from pure water boiling under a normal barometric pressure equal to 760 mm. of mercury. A variation of 1 mm. in the pressure will change the temperature 0.04°.

Zero is taken as the temperature at which pure ice melts when subject to an atmosphere of pressure. An increase of a whole atmosphere lowers the temperature of the melting point of ice 0.008°. This is to be distinguished from the effect of an atmosphere of pressure on the reading of a thermometer; by compressing the bulb it causes the thermometer to read about 0.°2 higher than if there was no pressure.

The fundamental distance on a normal thermometer is taken as the reading it would have at a true temperature of exactly 100°

minus the reading of its depressed freezing point. This, which should be exactly one hundred degrees, rarely is so. When the fundamental distance is taken in this way it is invariable with age. It is the same forty years after the thermometer is made as four hours after, provided the thermometer is kept at ordinary temperatures. The fundamental distance is not invariable when the raised freezing point is used in forming it instead of the depressed freezing point. In this case there is a constant diminution with age as the raised freezing point rises more rapidly than the boiling point. There is not uniformity of practice in the matter of forming the fundamental distance, but it is greatly to be desired.

Heating a thermometer to 350° causes a permanent increase in its fundamental distance between depressed freezing point and boiling, varying from 0.°4 to 0.°9. An increase of 0.°4 in the fundamental distance corresponds to a decrease of $\frac{1}{50}$ part in the coefficient of expansion of the glass.

The total correction of a mercurial normal thermometer for errors in its construction is composed of three parts:

1st. The correction for erroneous fundamental distance. For any temperature this is a proportional part of the difference between the fundamental distance and 100°.

2nd. The calibration correction. This is the correction to the scale marks considered as subdividing the capacity of the tube from 0° to 100° into one hundred equal parts. It involves variations in the bore of the tube as well as irregularities in the placing of the marks.

3rd. The correction at freezing point. This is the amount the thermometer reads in melting ice above or below 0°. At any time it is the observed reading of the thermometer in melting ice immediately after exposure to the temperature measured. Sometimes it is impossible and it is almost always inconvenient to observe the freezing point of the thermometer immediately after observing a temperature. In such a case the position of the depressed freezing point for that temperature must be computed from the law of the variation of the freezing point. It is always preferable however when the highest accuracy is required to actually observe the freezing point.

When a thermometer is put in ice this is what happens: The column falls rapidly at first, then more slowly. Presently it becomes stationary; finally it begins to rise. Fifteen minutes after the thermometer reaches its lowest reading this rise is about 0.°01. It is best in observing the freezing point of a thermometer to put it in ice thoroughly saturated with water. The ice should be contained in a vessel from which the water cannot flow off. A thermometer put in ice in this condition takes on the temperature of 0° more quickly than if put in dry ice containing a good deal of air. This method of observing the freezing point was introduced by Baudin of Paris.

There is another correction to a normal thermometer which is little known and rarely applied. Let:

$V =$ volume of bulb at 0°.

$v =$ volume of tube from 0° to 100° at temperature 0°.

$\beta =$ coefficient of cubical expansion of glass.

$\gamma =$ coefficient of expansion of mercury.

$T =$ thermometer-reading corrected for calibration, etc.

$t =$ true temperature.

A consideration of the construction of the thermometer leads to this equation:

$$V(1 + \beta t) + v(1 + \beta t)\frac{T}{100} = V(1 + \gamma t). \qquad (1)$$

The volume of the bulb at $t°$, plus the volume of that part of the tube corresponding to the thermometer reading, is equal to the volume of the mercury at $t°$. For $t = 100$ the equation becomes:

$$V(1 + \beta 100) + v(1 + \beta 100) = V(1 + \gamma 100). \qquad (2)$$

Eliminating $\frac{v}{V}$ from (2) by means of its value found from (1) we have:

$$T = t.\frac{1 + 100\beta}{1 + t\beta}. \qquad (3)$$

Taking as the coefficient of cubical expansion of glass, $\beta$, the quantity 0.000026, the following values are found for $T - t$, for the various readings of the thermometer from $- 40°$ to $+ 100°$.

| $T$ | $T - t$ |
|---|---|
| o c. | o c. |
| — 40. | —0.145 |
| — 20. | — 0.062 |
| 0. | 0.000 |
| + 20. | + 0.042 |
| 40. | + 0.062 |
| 50. | + 0.065 |
| 60. | + 0.062 |
| 80. | + 0.042 |
| 100. | 0.000 |

These are known as the Poggendorf corrections. They are due to the capacity of the tube from the zero to the one-hundred degree mark, being different at different temperatures.

In the following table are shown the results of the comparisons of a certain mercurial thermometer, Tounelot No. 4207, with the air-thermometer:

*Tounelot No. 4207.*

| Scale reading. | Correction as a normal thermometer. | Correction to reduce to the air-thermometer. | Differences. |
|---|---|---|---|
| o C. | o C. | o C. | o C. |
| 0.0 | 0.00 | 0.00 | 0.00 |
| 5.6 | + 0.05 | + 0.01 | + 0.04 |
| 11.1 | + 0.08 | 0.00 | + 0.08 |
| 16.1 | + 0.10 | + 0.02 | + 0.08 |
| 22.2 | + 0.15 | + 0.05 | + 0.10 |
| 25.2 | + 0.16 | + 0.07 | + 0.09 |
| 30.1. | + 0.20 | + 0.09 | + 0.11 |
| 35.4 | + 0.22 | + 0.10 | + 0.12 |
| 40.4 | + 0.25 | + 0.12 | + 0.13 |
| 45.1 | + 0.25 | + 0.16 | + 0.09 |
| 50.1 | + 0.24 | + 0.15 | + 0.09 |
| 52.7 | + 0.24 | + 0.16 | + 0.08 |
| 55.3 | + 0.24 | + 0.18 | + 0.06 |

If it be supposed that these differences are due to sensible terms in the expansion of glass and mercury dependent on the squares of the temperatures, an equation can be derived which will show that the maximum difference must be at 50°. But this is not so; the greatest difference is at about 40°. This agrees with what has been found by others. Rowland at Baltimore found the greatest difference at 40° to 45°; Mills in England found it at 35°, and Grunmach in Berlin at 30°.

Forming a theory of the differences on the supposition that they depend on the third powers of the temperatures as well as the squares, equation (4) is obtained, which gives the relation between the thermometer reading, $T$, and the true temperature, $t$.

$$T = t. \frac{1 + \beta_1 100}{1 + \beta_1 t} \cdot \frac{1 + \dfrac{\gamma_2 - \beta_2}{\gamma_1 - \beta_1} t + \dfrac{\gamma_3 - \beta_3}{\gamma_1 - \beta_1} t^2}{1 + \dfrac{\gamma_2 - \beta_2}{\gamma_1 - \beta_1} 100 + \dfrac{\gamma_3 - \beta_3}{\gamma_1 - \beta_1} (100)^2} \qquad (4)$$

This is only approximate. The effects of the second and third powers of the temperatures in changing the capacity of the tube from 0° to 100° are neglected. The capacity of the tube is only $\frac{1}{50}$ part of that of the bulb.

$\beta_1, \beta_2, \beta_3$ are the coefficients of expansion for glass for the first, second, and third powers of the temperature. $\gamma_1, \gamma_2, \gamma_3$ are the same for mercury.

To make an adjustment of the differences between the mercurial and air-thermometers equation (4) can be put in the form

$$(100t - t^2)x + (10000t - t^3)y + 0.000026t^2 - 0.0026t + \triangle = 0 \qquad (5)$$

in which $x = \dfrac{\gamma_2 - \beta_2}{\gamma_1 - \beta_1}, y = \dfrac{\gamma_3 - \beta_3}{\gamma_1 - \beta_1}$ and $\triangle = T - t$. This is less approximate than (4) but still sufficiently rigorous for the purpose intended.

Forming observation-equations on this model with the observed differences $\triangle$, as the absolute terms, and solving by the method of least squares, the values of $x$ and $y$ are found to be

$$x = -0.0001391$$
$$y = +0.000000863$$

Substituting these in (5) it becomes

$$-0.00788t + 0.000165t^2 - 0.000000863t^3 + \triangle = 0. \qquad (6)$$

Differentiating (6) with respect to $t$ and $\triangle$, and putting $\dfrac{dt}{d\triangle} = 0$, the following quadratic-equation is found

$$0.000002589t^2 - 0.000330t + 0.00788 = 0, \qquad (7)$$

the solution of which gives for the temperatures at which the differences between the mercurial and air-thermometer are greatest $t = 31.8$ and $t = 95.8$. To find the temperatures at which the mercurial and air-thermometer agree, put $\triangle = 0$ in (6); the values of $t$ that then satisfy the equation are $t = 0$, $t = 100$, and $t = 91$.

At 32° the mercurial thermometer reads higher than the air-thermometer, at 96°, it reads lower. A curve representing the differences has the following form:

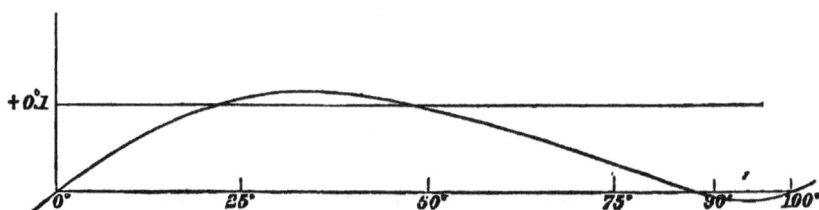

Fig. 1.—$T$ 4207 *minus* Air-thermometer. Abscissas = Temperatures. Centigrade. Ordinates = Differences.

This agrees with what has been observed by Dr. Grunmach of the Berlin Aichungs Commission. He found the maximum difference on a certain thermometer to be $+ 0.°12$ at $29.°8$, and another secondary maximum of $- 0.°04$ at $82°$.

A slight change in the values of $x$ and $y$ will make a large change in the position of the secondary maximum. For another thermometer investigated at the Signal Office this point was found to be at $130°$.

The values of $x$ and $y$ can be analyzed still further to ascertain whether consistent with known physical properties of glass and mercury.

Taking Broch's values for the expansion of mercury, which are based on a re-reduction of Regnault's observations, it is found, adapting the figures to the notation used here, that:

$$\gamma_1 = + 0.000181792$$
$$\gamma_2 = + 0.000000000175$$
$$\gamma_3 = + 0.000000000035116$$

Substituting these values of $\gamma_1$ and $\gamma_2$ in the equations

$$x = \frac{\gamma_2 - \beta_2}{\gamma_1 - \beta_1}, \quad y = \frac{\gamma_3 - \beta_3}{\gamma_1 - \beta_1}$$

and aking $\beta = 0.000026$ we get

$$\beta_2 = + 0.000000021859$$
$$\beta_3 = + 0.000000000099512$$

The linear coefficient of expansion of a specimen of glass, such as is used in barometer-tubes and thermometers, has been very carefully determined by Dr. Benoit of the International Bureau of Weights and Measures. Deriving from this the cubical coefficient, in the notation used here, we have Dr. Benoit's values

$$\beta_1 = + 0.0000252$$
$$\beta_2 = + 0.0000000144$$

It will thus be seen that there is a good agreement between the two values of $\beta_2$ found by the two different processes.

Remarks were made by Messrs. GILBERT, HARKNESS, and PAUL.

Mr. J. H. KIDDER made a communication on

## THE GILDING OF THERMOMETER BULBS,

and was followed by Mr. H. ALLEN HAZEN on

## EFFECTS OF SOLAR RADIATION UPON THERMOMETER BULBS HAVING DIFFERENT METALLIC COVERINGS.

[Abstract.]

After showing the importance of shielding from or measuring the effects of solar radiation upon thermometer bulbs used in determining the air temperature and indicating the attempts that have been made in obtaining satisfactory results in the past, Mr. HAZEN explained the most recent work of Prof. Wild, of St. Petersburg. Prof. Wild had a bulb coated with copper by the galvanoplastic method, then the copper with gold, which latter was highly polished. From theoretical considerations he established a formula as follows :

$$t_a = t_m - c\,(t_s - t_m),$$

in which $t_a$ = temperature of air sought; $t_m$ = that of the metallic-covered bulb; $t_s$ = that of a black bulb; and $c$ is a constant which he assumed at .15. The published observations indicated a very high reading of the metal-covered bulb, even higher than would have been obtained with a bare glass bulb.

Mr. HAZEN's own experiments consisted in comparisons between bulbs as follows: (1) black, (2) bright, (3, 4) with silver and gold deposited in an exceedingly thin layer and giving a most admirable surface, and (5, 6) with silver and gold deposited on copper as nearly as possible as suggested by Prof. WILD. These thermometers were exposed in sunshine in the open air as well as indoors. It was very difficult to shield from other radiations, dark heat, etc., but it was found that an exposure of two feet from a window pane gave fairly good results. The results are given in the accompanying table. In computing the actual value of $t_a$ from the formula it was found necessary to take the readings of the black and bright bulbs at each set and compute the $t_a$ by using .6 as the constant. It was found that the black and bright bulbs were the only ones that did not deteriorate from day to day. All the metal-covered bulbs were carefully polished each day.

*Comparison of various coverings of thermometer bulbs (out of doors).*

| Date. | Black. | Bright lower than black. | | Silver deposit. | | Gold deposit. | | Silver plated. | | Gold plated. | |
|---|---|---|---|---|---|---|---|---|---|---|---|
| | ° | ° | c. | ° | c. | ° | c. | ° | c | ° | c. |
| Apr. 14, '86........ | 87.5 | 7.2 | .60 | 9.0 | .29 | 7.0 | .65 | 7.7 | .49 | 5.8 | 1.00 |
| Apr. 19 ........ | 76.3 | 6.5 | .60 | 8.2 | .28 | 6.3 | .65 | 6.9 | .50 | 5.1 | 1.04 |
| " ........ | 89.0 | 4.9 | .60 | 6.3 | .24 | 4.7 | .67 | 5.5 | .42 | 3.7 | 1.11 |
| Apr. 20 ........ | 76.8 | 6.0 | .60 | 7.3 | .32 | 5.6 | .71 | 6.3 | .52 | 5.1 | .88 |
| " ........ | 83.9 | 7.3 | .60 | 9.2 | .27 | 6.7 | .75 | 7.4 | .50 | 5.8 | 1.02 |
| " .. ..... | 89.3 | 5.6 | .60 | 7.6 | .18 | 5.8 | .55 | 6.1 | .48 | 5.0 | .80 |
| May 3 .. ..... | 76.9 | 7.3 | .60 | 9.4 | .24 | 6.9 | .70 | 7.7 | .52 | 6.3 | .86 |
| " ........ | 82.8 | 6.9 | .60 | 8.8 | .25 | 6.6 | .67 | 7.7 | .43 | 5.8 | .89 |
| " ........ | 92.1 | 8.3 | .60 | 10.3 | .27 | 7.9 | .66 | 8.2 | .60 | 7.6 | .72 |
| Mean ............ | 83.8 | 6.7 | .60 | 8.5 | .26 | 6.4 | .67 | 7.1 | .50 | 5.6 | .81 |

The results in this table may be taken as relatively accurate, though the absolute values cannot be relied on. We find the constant in the case of the silver-deposited bulb less than one half that of the bright or bare glass bulb; the gold-deposited runs a little higher; the silver-plated somewhat lower, while the gold-plated is the highest of all. In the gold-deposited thermometer the covering was so thin that it showed a green light through. It has been shown that under these circumstances gold transmits a little heat. The gold-plated thermometer gives a remarkable result due partly to the thickness of the metal and partly to a slight roughness, it having been found impossible to deposit a perfectly smooth copper surface as a base.

Incidentally these plated bulbs have furnished corroboration of a point presented to the Society some time ago, namely, the effect of a layer of ice upon a bulb in contracting it at low temperatures. It is plain that at the temperature of deposition there would be no effect, but as the temperature was lowered, since the metal contracts faster than glass, the tendency would be to a too high reading of the thermometer—*e. g.*, Thermometer No. 840 gave too high readings at different temperatures as follows: at 12°, 3.6; 30°, 1.4; 40°, 1.0; 50°, .7; 60°, .4, and 70°, .1.

Remarks on these communications were made by Messrs. PAUL, MENDENHALL, BILLINGS, and WOODWARD.

288TH MEETING.                                    MAY 22, 1886.

Vice-President HARKNESS in the Chair.

Forty-eight members and visitors present.

Mr. NEWTON L. BATES read a communication entitled

ORGANIC CELLS OF UNKNOWN ORIGIN AND FORM FOUND IN
HUMAN FÆCES (TWO CASES),*

illustrating his remarks by specimens exhibited under the microscope.

Referring to the composition of these cells, reported to contain four
per cent. of silica, Mr. CLARKE called attention to the desirability
of testing for other inorganic substances. In reply to a question as
to where in the intestinal canal the cells had been found, Mr. BATES
replied that the evidence in the case indicated the lower part of
the canal.

Mr. J. S. BILLINGS made a communication

ON MUSEUM SPECIMENS ILLUSTRATING BIOLOGY.

[Abstract.]

After referring to the increasing interest on the part of the public in Biology and Natural History, as theories of evolution, natural
selection, etc., are becoming better understood, and to the consequent increasing importance of museums or parts of museums intended to illustrate the structure and functions of animals, several
types of museums were briefly described, including the collection at
Florence, the Hunterian Museum, the average Medical School Museum, the museums proposed by Drs. Roberts, Wilder, and Shu-
feldt, etc.

The special scope and purposes of the Army Medical Museum
were stated to be to illustrate: 1. The effects, immediate and remote,
of wounds and of those diseases most prevalent in the army, i. e.,

---

* These cells are now believed to have come from the banana. If a transverse slice of banana is placed in a watch glass and covered with strong
nitric acid, two rows of cells become colored and are apparent to the naked
eye. A note on the subject may be found in the Boston Medical and Surgi-
cal Journal, March 10th, 1883, p. 446. Similar changes doubtless take
place by prolonged stay in the intestine, and when found clean and free
from vegetable fiber these cells are likely to deceive expert microscopists.

the diseases and injuries of adult males.  2. The work of the medical department of an army, modes of transporting sick and wounded, hospitals, medical supplies, instruments, etc.  3. Human anatomy and pathology of both sexes and all ages.  This requires many specimens in comparative anatomy and pathology, which are indispensable for a correct understanding of the structure, development, abnormalities and diseases of man.  It is not however proposed to form a museum of comparative anatomy—that belongs to the functions of the National Museum.  4. To show the morphological basis, or want of such basis, for ethnological classification, more especially of the native races of America.  This includes anthropometry and craniology.  5. To illustrate for medical investigators and teachers the latest methods, the newest apparatus, etc., for biological investigation, and various modes of preparing and mounting specimens.  In connection with this it is hoped to induce original workers to deposit in the museum type specimens or series of specimens illustrating their discoveries and methods.

The classification and arrangement of specimens which it is proposed to carry out in the new Museum building were then briefly described.

Various modes of preparing specimens were shown, including dissections under spirit, frozen and tinted sections, injections by corrosion, etc., and the difficulties in making and preserving such specimens were explained.

A second communication on the same subject was made by Mr. G. BROWN GOODE, and a third, by request, by Mr. FREDERICK A. LUCAS, of the National Museum, who spoke more especially of osteological specimens.  The papers were all fully illustrated by specimens from the Medical and National Museums.

Mr. GEORGE P. MERRILL made a communication

### ON GEOLOGICAL MUSEUMS.

Remarks were made on the general subject of museums and their management by Prof. EDWARD MORSE, of Salem, Mass.

289TH MEETING.                                          OCTOBER 9, 1886.

The President in the Chair.

Seventy-eight members and guests present.

The Chair announced the election to membership of Messrs. COOPER CURTICE, HENRY MITCHELL, HENRY GUSTAV BEYER, and NEWTON LEMUEL BATES.

The subject for the evening was

### THE CHARLESTON EARTHQUAKE,

which was discussed by Messrs. T. C. MENDENHALL and W J Mc-GEE and Prof. CHARLES G. ROCKWOOD of Princeton, N. J.

Mr. MENDENHALL spoke of the odor observed on Sullivan's Island previous to the shock, of the time of the clock-stopping shock in Charleston, of the detonations accompanying the various shocks and heard in Charleston, Summerville, and elsewhere, of the torsional movements of monuments, and of the directions in which various structures were thrown down. He exhibited an isoseismal map compiled from data gathered by the U. S. Signal Service. He also stated at second hand a novel theory for the origin of the earthquake, and spoke of the convergence of opinion of geologists and physicists in regard to the condition of the interior of the earth. He accepted as the time of the clock-stopping shock in Charleston 9h. 51′ 20″ P. M.

Mr. McGEE described the geological relations of Charleston, showing that a depth of 2,000 feet of clastic rocks had been demonstrated beneath the city, and that the total depth to the crystalline rocks might be as much as one mile. He described the phenomena of deformation of rails and other railway structures, gave in detail an observation of a severe shock at Summerville, described the detonations, and exhibited numerous photographs illustrating the destructive work of the earthquake and the formation of craterlets and sinks.

Prof. ROCKWOOD exhibited an isoseismal map compiled from data gathered by the Earthquake Commission, and characterized by far greater irregularities in the form of contours than were shown by Mr. Mendenhall's map. He called attention to the fact that nearly all the isoseismal curves show salient angles toward the northwest. He dwelt upon the exceptional nature of this opportunity for earthquake investigation, and urged that the utmost advantage be taken of it.  He spoke also of the complexity of earthquake movements, and the difficulties to be overcome in analysing them.  He thought that the conspicious inequality in the violence of the shocks at localities not widely separated was to be ascribed to the intersection and combination of rock waves deflected by reflection and refraction.  Two intersecting waves would be especially destructive at their nodal points, and comparatively harmless at their points of interference.

It was announced by the President that the discussion would be continued at the next meeting.

290TH MEETING.                                    OCTOBER 23, 1886.

The President in the Chair.

Sixty-seven members and guests present.

The discussion of

### THE CHARLESTON EARTHQUAKE

was resumed, the principal speakers being Mr. EVERETT HAYDEN and Mr. H. M. PAUL.  Remarks were made by Messrs. McGEE, BILLINGS, ROBINSON, DUTTON, BELL, CLARKE, and GILBERT, and by Dr. E. P. HOWLAND, who was present by invitation.

Mr. HAYDEN first discussed a chart on which were plotted the areas disturbed by earthquakes in the southeastern United States from 1874 to 1885, compiled from Rockwood's Notes in the American Journal of Science.  This indicated two earthquake belts, one along the Appalachians, the other along the coast.  In the former 28 shocks are recorded, in only 3 of which the area is as great as 1,000 square miles (2 in central Virginia, and 1 in western North Carolina and northern Georgia); in the latter 5 are recorded, only 1 of which (that of 1879 in Florida and Georgia) was at all severe.  So far as this evidence goes, therefore, we should have expected a severe shock like that of August 31st to have originated in the Appalachians and to have been orogenic in character, an accompaniment of the gradual elevation of the range.

He then proceeded to give a summary of the information which had reached the Geological Survey up to date, illustrating by charts of isoseismal and coseismal lines.  The geologic and physical phenomena in the region of greatest intensity having been already dis-

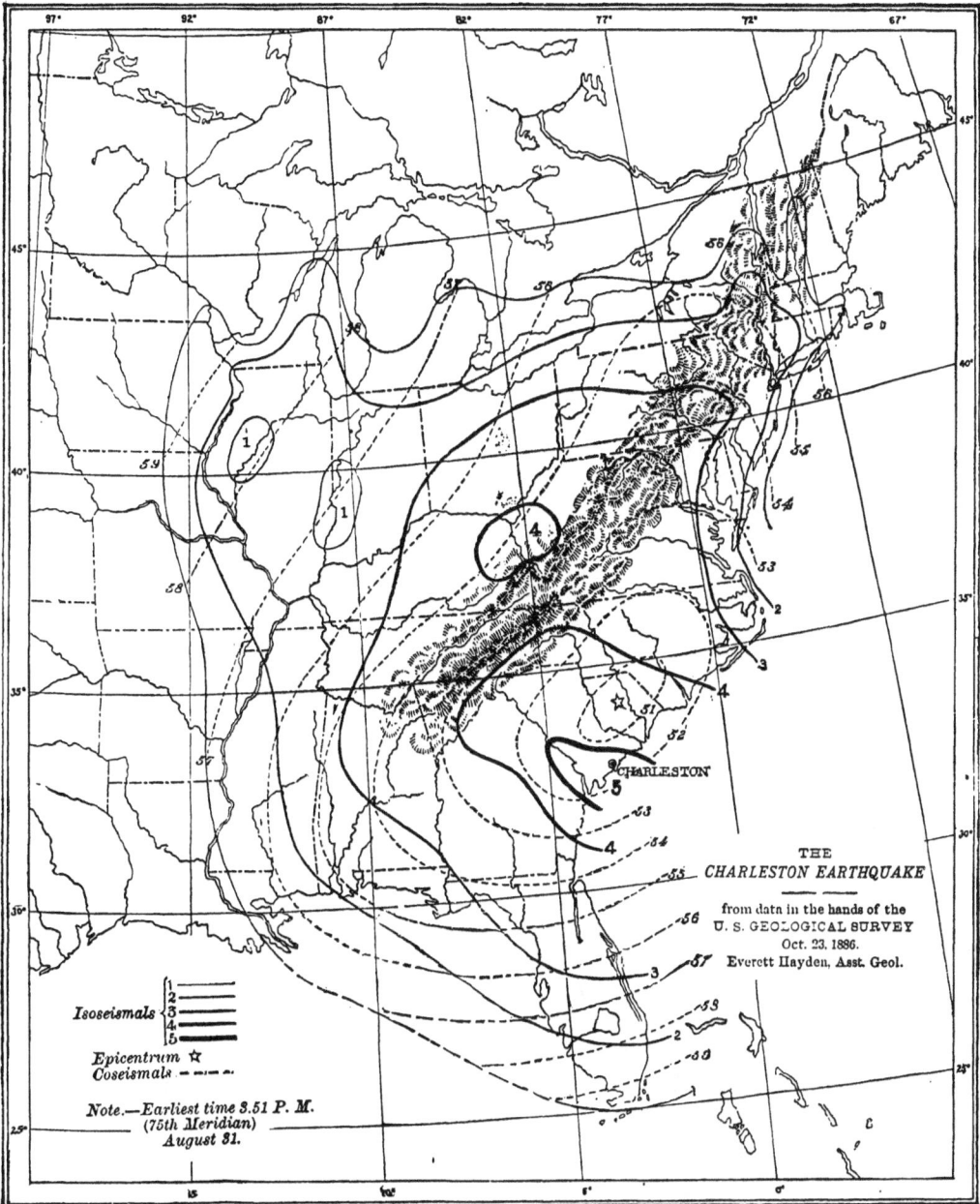

THE
CHARLESTON EARTHQUAKE
from data in the hands of the
U. S. GEOLOGICAL SURVEY
Oct. 23, 1886.
Everett Hayden, Asst. Geol.

Isoseismals {1 2 3 4 5
Epicentrum ☆
Coseismals

Note.—Earliest time 3.51 P. M.
(75th Meridian)
August 31.

cussed at the last meeting were not touched upon. The data used came for the most part from correspondence with private parties, although some valuable information had been received from the Signal Service, Light-House Board, and Hydrographic Office—only a small portion, however, of what they would eventually furnish. The State Department had reported that the shock was felt very slightly in Bermuda, and would report later as to Cuba and the Bahamas.

The *isoseismals* were plotted on an enlarged photograph of the relief model of the eastern United States in the Coast Survey Office, which illustrated the general topographic features more graphically than an ordinary map. In the discussion special attention was paid to explaining their irregular shapes by reference to the surface configuration as well as the geologic structure of the country. The inclosed area, marked 4, in West Virginia and Kentucky, well illustrates the fact that a shock may be felt with greater severity beyond a mountain range than in its midst. The similar isolated areas of less intensity in Indiana and Illinois are also typical of variations due to local conformations; were it possible to plot intensities in still greater detail there would doubtless be hundreds of such isolated districts all through the disturbed area. Other points dwelt upon were the unobstructed transmission of the vibrations along the parallel ridges of the range and up the valleys of the Connecticut and Hudson rivers; the obstruction offered by ridges, valleys, and strike of strata transverse to the direction of propagation; and the rapid loss of energy in the sands and alluvial deposits of the northeast coast and lower Mississippi valley. The total land area included within the outer isoseismal is 774,000 square miles, and if we add only half as much more for ocean area it closely approximates to that given by Reclus for the great Lisbon earthquake. Special acknowledgement was made for valuable positive and negative reports received from members of the New England Meteorological Society through their secretary, Prof. W. M. Davis; they accurately fixed the limits of the disturbed area in New England.

The *coseismals* were plotted from the most reliable and consistent among a hundred or more good time observations, and special care was taken to make them conform to the actual facts, uninfluenced by any preconceived theoretical ideas. Attention was called to certain peculiarities of these lines, such as their noticeable prolongation to the northward and southwestward; the narrow intervals where they run along the western flanks of the Appalachians and across the Florida peninsula; and the wide intervals in the Ohio and upper Mississippi valleys. The following considerations were offered as helping to explain these apparently anomalous features: The crystalline Archean rocks and parallel ridges of the mountain chain favor a rapid axial transmission of the vibrations. The first tremors spreading to the northwestward, however, are cut off and deadened by the mountains, so that it is a later phase of the wave which is felt and recorded beyond; having passed the range it then spreads with little obstruction and high velocity through comparatively level strata. Similarly in the littoral and alluvial deposits to the northeast, south, and southwest the earlier tremors are lost and later phases of the wave are successively recorded. It is especially to be remembered that all these times are from non-instrumental observations; an exact instrumental observation made at Toronto, Ontario, by Prof. Chas. Carpmael (9: 54: 50 P. M.), could not be used here because so early as to be wholly inconsistent with all other reliable but non-instrumental observations.

The epicentrum, or point on the surface directly above that part of the deep-lying fissure where the earliest vibrations originated, is placed by these coseismals about a hundred miles north of Charleston, which is not at all inconsistent with the fact that the greatest damage was caused in that city. In fact, it is to be expected that the destruction of buildings should be greater at a distance, where the angle of emergence is less. Moreover, most of the evidence seems to point to a very deep-lying origin, in which case one cannot but attribute much of the local damage, as well as the continuance of shocks of considerable intensity but small area, to the character of the recent geologic formations in that region. Borings for artesian wells at Charleston indicate that the Tertiary and Cretaceous strata are very heterogeneous in character—sands, clays, limestone-marls, and imperfectly consolidated beds of conglomerate, with occasional cavities containing running water—and much of the city is built upon made land. Such considerations may explain the extremely local character of many of the shocks which have been felt at various points in the State of South Carolina since August 31st, as well as the great intensity of the shock at Charleston on that date. Local sinks in the ground are reported even in northern Florida, far from the origin of the disturbance. An analogous case is seen in the fact that the great destruction of life and property at Lisbon, in 1755, occurred in those portions of the city built on weak Tertiary formations, while houses on the firmly consolidated Secondary rocks suffered little damage.

The following velocities of wave transmission are indicated :

| To— | Feet per second. | Miles per minute. |
|---|---|---|
| Toronto, Ontario | 15,000 | 170 |
| Washington, D. C. | 13,000 | 148 |
| Prairie du Chien, Wisc. | 9,300 | 106 |
| Average | 12,400 | 141 |

By way of comparison the following recorded velocities are of interest: Lisbon, 1755, 2,000 feet per second; Naples, 1857, 1775; St. Lawrence valley, 1870, 12,000 ; England, 1884, 9,200.

Reported directions of transmission, while very often what we should expect, are yet generally so contradictory as to be of little value. Similarly the number of shocks felt is recorded so differently by different observers as to be very confusing ; the occurrence of two shocks at a point at some distance from the origin is explained by the hypothesis that the first traveled rapidly through the hard under-lying rocks, and the second more slowly through the softer and more recent strata above. The shock is reported as accompanied by sounds of greater or less intensity all along the extent of Archean rocks from northern Alabama to Connecticut, and at points on the coastal plain within a radius of about 300 miles from the origin.

The coincidence of an unusually high tide is worthy of remark ; the moon was near perigee, and there had been an eclipse of the sun only three days previously. The fact that no sea-wave was caused by the shock confirms the conclusion that the origin was inland. The weather is generally reported to have been very still and sultry, although there were no unusual barometric conditions. The summer had been an unusually dry one.

Only two other recorded earthquakes in North America can be compared with this in either area or intensity: that at New Madrid, Missouri, in 1811, which was probably fully its equal, and that in the St. Lawrence valley in 1870, equal possibly in area but not in intensity.

Mr. PAUL explained the generally received classification of earthquake waves, described some of the results of the Tokio earthquake studies, discussed the relation of destructiveness to amplitude of vibration and to rate of acceleration, and requested Mr. McGee to describe more specifically than at the last meeting the phenomena observed by him in connection with a severe shock in Summerville.

Mr. McGee said that during the tremor the bedstead upon which he lay left the floor from fifteen to thirty times, the departures being roughly estimated at three per second. The clear ascent, as judged from the force of the return blow, ranged from one-fourth inch to two or three inches. The bed stood on the ground floor of a wooden house, supported on piers. Earlier shocks had crushed or driven down the piers under the heavier parts of the house, so that the weight was borne in large part by piers under verandas, etc.

Mr. PAUL said that if the floor in descending separated from the legs of the bedstead its acceleration of motion must exceed that due to gravitation. The fall of the earth manifestly could not be faster than that of the bedstead. Assuming the accuracy of the observation the only possible explanation would seem to be that the floor had behaved as an elastic spring.

Dr. HOWLAND described a shock observed by himself, and said that of some hundreds of chimneys observed by him in Charleston 75 per cent. had fallen to the west.

The PRESIDENT spoke of the bearing of the earthquake upon sanitary matters. The water mains in Charleston had been run in some places through sewers, and there is no assurance that these mains are now in such condition as to render contamination impossible.

Mr. DUTTON inferred from the magnitude of the area through which the shock was felt, as compared with its moderate destructiveness in the central region, that the centrum lay at great depth.

Mr. BELL was much interested in the statement that, even at considerable distances from the centre of disturbance, the noise accompanying the earthquake either preceded the shock or was perceived simultaneously with it. This, he thought, indicated that the sound was of local origin. The great velocity with which the earthquake disturbance had been propagated seemed to him to preclude the idea of a sound-wave from the centre of disturbance as the cause of the noise perceived. Any sound due to this cause should, he thought, at considerable distances, have been observed after the experience of the shock.

He also spoke of the worthlessness of testimony regarding the direction of sounds proceeding from points below the observer. In this connection he directed attention to experiments relating to Binaural Audition which he had communicated to the American Association for the Advancement of Science in 1879;* and he recommended any one who placed reliance on the testimony that had been adduced relating to the direction of the earthquake noises to try the following experiment:

To one end of a long pole attach an ordinary electric call-bell, and at the other end place a push-knob, by means of which the experimenter can at will ring the bell. Let the person whose credibility as a witness is to be tested stand with his feet considerably apart, with his eyes closed and head still. The experimenter can then silently move the call-bell into any desired position before ringing, and the observer can indicate his appreciation of the direction of the sound by pointing to the place from whence he conceives it to have emanated,

Such being the disposition of the parties the experiment Mr. Bell would recommend is the following: Carefully and silently introduce the end of the pole between the legs of the observer so that the bell is directly underneath him. Now ring and ask your witness to indicate by pointing the position of the bell. He had tried the experiment many times, and had been surprised, and even startled, by the result. The observer usually formed a distinct judgment as to the direction of the sound, but the one feature that was common to all the experiments was *that the indicated direction was wrong.*

Mr. Gilbert remarked that the simultaneous occurrence of detonations and tremors indicated that the sound waves were identical with some at least of the waves constituting the earthquake. It was therefore legitimate to compare the velocity of transmission of the earthquake waves with the velocity of transmission of sound in various media, and such comparison indicates that the portion of the crust traversed by the earthquake waves was characterized by an elasticity between that of gold and that of iron.

Mr. Bell thought that deductions based upon the assumption that the disturbance had been propagated with the velocity of

---

* The paper was published *in extenso* in the American Journal of Otology for July, 1880, vol. II, p. 169. See also Nature, vol. XXI, p. 310, and vol. XXII, pp. 586–7.

sound should be received with caution; for it is well known that very great and sudden disturbances may be propagated through a medium with a greater velocity than the normal velocity of sound in that medium. In Captain Parry's Arctic expedition it was noticed that distant observers heard the report of cannon before hearing the command to fire.

The PRESIDENT announced that Mr. T. C. MENDENHALL, having removed from the city, had resigned his position on the General Committee of the Society, and that the committee had filled the vacancy by the election of Mr. WILLIAM B. TAYLOR.

---

291ST MEETING.                                   NOVEMBER 6, 1886.

Vice-President MALLERY in the Chair.

Twenty-three members and guests present.

Mr. O. T. MASON made a communication on

### BOWYERS AND FLETCHERS.

[Abstract.]

Whatever may be our theory of creation, the arts of mankind proceed from the same sources as the genera and species of natural objects. The design of this paper is to demonstrate, by means of an art almost universally dispersed in time and place, that we may regard the implements and products of human industry in the light of biological specimens. They may be divided into families, genera, and species. They may be studied in their several ontogenies (that is, we may watch the unfolding of each individual thing from its raw material to its finished production). They may be regarded as the products of specific evolution out of natural objects serving human wants and up to the most delicate machine performing the same function. They may be modified by their relationship, one to another, in sets, outfits, apparatus, just as the insect and flower are co-ordinately transformed. They observe the law of change under environment and geographical distribution.

The bow, at first, was only an elastic limb or branch transformed little or none at all. From this parent form have developed three types under the control of the material, namely, the perfect, simple bow, in lands where elastic woods abound; the compound bow, in localities where, by choice or necessity, horn, bone, and antler are preferred as material; and the sinew-backed bow, with its two sub-types of the corded back and the solid back. Each one of these types may be subdivided indefinitely by ethnic marks.

The arrow, at first a reed or twig unmodified, was only a shaft with merely an indication of a head as in some of the lower forms. From this, by a normal evolution, have come the feather, fore-shaft, head, and barbs, differentiating into endless varieties under the stress of material, definite functions, and the thousand and one forces which together we may call its environment.

A large collection of bows and arrows varying in material, form, and origin was exhibited to exemplify the theory set forth.

The communication was discussed by Messrs. TAYLOR, E. FAR-QUHAR, RILEY, HARKNESS, DALL, and ELLIOTT. Mr. Taylor called attention to the break in the evolutionary history of the bow at its very beginning. The stride from the elastic throwing-stick to the bow is immense. The discussion turned chiefly upon the proper basis for museum classification of ethnological material, Messrs. Riley, Dall, and Elliott advocating a classification primarily by peoples or races, and Mr. Mason defending the evolutionary system, where classification by races is supplemented and traversed by a classification in which articles of a kind are placed together.

Mr. G. K. GILBERT began a communication on

CERTAIN NEW AND SMALL MOUNTAIN RANGES,

which was unfinished when the hour for adjournment arrived.

---

292D MEETING.                                    NOVEMBER 20, 1886.

The President in the Chair.

Forty-two members and guests present.

Mr. G. K. GILBERT completed his communication

ON CERTAIN NEW AND SMALL MOUNTAIN RANGES,

and remarks were made by Messrs. BILLINGS and HAZEN.

Mr. THOMAS RUSSELL presented a communication on

NORMAL BAROMETERS,

which was discussed by Messrs. BILLINGS and HARKNESS.

Mr. N. H. DARTON read a paper

ON THE OCCURRENCE OF COPPER ORE IN THE TRIAS OF THE EASTERN UNITED STATES,

and Mr. J. S. DILLER followed with a communication on

THE LATEST VOLCANIC ERUPTION IN NORTHERN CALIFORNIA AND ITS PECULIAR LAVA.

[Published in Am. Jour. Sci., 3d series, vol. xxxiii.]

Remarks on the last paper were made by Mr. IDDINGS.

---

293D MEETING.　　　　　　　　　　　　DECEMBER 4, 1886.

By courtesy of the trustees of the Columbian University the meeting was held in the law lecture-room of the University building. Invitation to attend the meeting was extended to the members of the Anthropological, Biological, and Chemical societies and of the Cosmos Club. Two hundred and two persons were present.

Vice-President MALLERY presided.

The Chair read a letter from the secretary of the Chemical Society inviting the members of the Philosophical Society to listen, on the evening of Dec. 9, to an address by Prof. H. W. WILEY, retiring president of the Chemical Society, on "Our Sugar Supply."

President BILLINGS then presented his annual address, the subject being

SCIENTIFIC MEN AND THEIR DUTIES.

[Printed in full on pp. XXXV–LVI of this volume.]

A vote of thanks for the address was passed by the audience.

294TH MEETING. DECEMBER 18, 1886.

### THE SIXTEENTH ANNUAL MEETING.

The President in the Chair.

The minutes of the 292d and 293d meetings were read and approved.

The Chair read the order of business as prescribed by the Standing Rules.

The Secretary read the minutes of the fifteenth annual meeting.

The Chair announced the election to membership of Messrs. JOSEPH CLAYBAUGH GORDON, NELSON HORATIO DARTON, MARSHALL McDONALD, WILLIAM LEE TRENHOLM, and WILLIAM FRANCIS HILLEBRAND.

The annual report of the Secretaries was read. [See page XXIX.]

The annual report of the Treasurer was read [see page XXX] and referred to an Auditing Committee, consisting of Messrs. R. S. Woodward, S. M. Burnett, and J. H. Kidder.

The Treasurer read the list of members entitled, under Standing Rule 14, to vote for officers.

After a recess of five minutes, the Society proceeded to the election of officers for the year 1887. [The result of the election appears on page XV.]

The following amendment to the Constitution was offered by Mr. W. H. Dall, and laid on the table, as required by Article VI of the Constitution.

In Article III insert, after " consisting of," the words *the ex-presidents of the Society.*

In Article IV insert, before " other members," the word *nine.* [The effect of the amendment is to make ex-presidents of the Society permanent members of the General Committee.]

The rough minutes of the meeting were then read, and the Society adjourned.

# BULLETIN

OF THE

# PHILOSOPHICAL SOCIETY OF WASHINGTON.

---

## MATHEMATICAL SECTION.

# STANDING RULES

OF THE

# MATHEMATICAL SECTION.

1. The object of this Section is the consideration and discussion of papers relating to pure or applied mathematics.

2. The special officers of the Section shall be a Chairman and a Secretary, who shall be elected at the first meeting of the Section in each year, and discharge the duties usually attaching to those offices.

3. To bring a paper regularly before the Section it must be submitted to the Standing Committee on Communications for the stated meetings of the Society, with the statement that it is for the Mathematical Section.

4. Meetings shall be called by the Standing Committee on Communications whenever the extent or importance of the papers submitted and approved appear to justify it.

5. All members of the Philosophical Society who wish to do so may take part in the meetings of this Section.

6. To every member who shall have notified the Secretary of the General Committee of his desire to receive them, announcements of the meetings of the Sections shall be sent by mail.

7. The Section shall have power to adopt such rules of procedure as it may find expedient.

# OFFICERS

OF THE

# MATHEMATICAL SECTION FOR 1886.

*Chairman,* WM. B. TAYLOR.          *Secretary,* MARCUS BAKER.

## LIST OF MEMBERS WHO RECEIVE ANNOUNCEMENT OF THE MEETINGS.

| | |
|---|---|
| ABBE, C. | HARKNESS, W. |
| AVERY, R. S. | HAZEN, H. A. |
| BAKER, M. | HILL, G. W. |
| BATES, H. H. | HODGKINS, H. L. |
| BILLINGS, J. S. | KING, A. F. A. |
| BURGESS, E. S. | KUMMEL, C. H. |
| CHRISTIE, A. S. | LEFAVOUR, E. B. |
| COFFIN, J. H. C. | McGEE, W J |
| CURTIS, G. E. | MARTIN, A. |
| DeLAND, T. L. | NEWCOMB, S. |
| DOOLITTLE, M. H. | PAUL, H. M. |
| EASTMAN, J. R. | RITTER, W. F. M'K. |
| EIMBECK, W. | ROBINSON, T. |
| ELLIOTT, E. B. | SMILEY, C. W. |
| FARQUHAR, H. | STONE, O. |
| FLINT, A. S. | TAYLOR, W. B. |
| GILBERT, G. K. | UPTON, W. W. |
| GORE, J. H. | WINLOCK, W. C. |
| GREEN, B. R. | WOODWARD, R. S. |
| HALL, A. | ZIWET, A. |

52

# BULLETIN

OF THE

# MATHEMATICAL SECTION.

22D MEETING.                                    MARCH 24, 1886.

The Chairman, Mr. G. W. HILL, presided.

Present, fourteen members and one invited guest.

Election of officers of the section for the year 1886 was conducted by ballot, and resulted in the choice of Mr. W. B. TAYLOR as chairman and Mr. MARCUS BAKER as secretary.

Mr. HENRY FARQUHAR made a communication on

### A COMPARISON OF THE BOSS AND AUWERS DECLINATION STANDARDS,

in which the systematic difference between the standard of Dr. Auwers' "Fundamental-Catalog" and the "Normal System of Prof. Boss was considered as a function of declination, and shown to be almost completely explained by the supposition of a tube-flexure inadequately allowed for, in the observations on which one or other standard depends.

[A paper covering the same ground was afterward read by Mr. FARQUHAR before the American Association for the Advancement of Science, and an abstract of it will appear in the proceedings of that body for the Buffalo meeting, 1886.]

The paper was briefly discussed by Messrs. HALL, HILL, and WOODWARD.

Mr. R. S. WOODWARD then read a paper entitled

### ON THE POSITION AND SHAPE OF THE GEOID AS DEPENDENT ON LOCAL MASSES.

[This will be published as a Bulletin of the United States Geological Survey.]

53

23D MEETING.                                    APRIL 14, 1886.

The Chairman, Mr. WILLIAM B. TAYLOR, elected at the last meeting, presided, and upon taking the chair offered a few remarks expressive of his appreciation of the honor conferred upon him and of his desire for the prosperity of the Section.

Present, seventeen members and one guest.

Mr. R. S. WOODWARD recited, in abstract, the principal points of his paper of the preceding meeting, and there ensued a general discussion of the subject, in which Messrs. BAKER, DOOLITTLE, HILL, PAUL, THOMAS RUSSELL, STONE, TAYLOR, AND WOODWARD participated.

Mr. C. H. KUMMELL then presented a paper

ON THE USE OF SOMOFF'S THEOREM FOR EVALUATION OF THE ELLIPTIC INTEGRAL OF THE THIRD SPECIES.

[This paper will appear in full in Annals of Mathematics, vol. 2, Nos. 4 and 5.]

Remarks were made by Mr. HILL.

# INDEX.

# BULLETINS

OF THE

## PHILOSOPHICAL SOCIETY ,OF WASHINGTON.

---

Vol. I, March, 1871, to June, 1874.  8vo, pp. 158.  Price $1.

Vol. II, October 10, 1874, to November 2, 1878.  Contains two memoirs of the late Joseph Henry, a portrait, two lithographs, and a map.  8vo, pp. 452.  Price $2.50.

Vol. III, November 9, 1878, to June 19, 1880.  8vo, pp. 169. Price $1.

Vol IV, October 9, 1880, to June 11, 1881.  8vo, pp. 189. Price $1.

Vol. V, October 8, 1881, to December 16, 1882.  8vo, pp. 189. Price $1.

Vol. VI, containing the minutes of the Society for the year 1883, and the minutes of the Mathematical Section from its organization, March 29, to the close of the year.  8vo, pp. 225.  Price $1.

Vol. VII, containing the minutes of the Society and of the Mathematical Section for the year 1884.  8vo, pp. 195.  Price $1.

Vol. VIII, containing the minutes of the Society and of the Mathematical Section for the year 1885.  8vo, pp. 110.  Price $0.75.

Vol. IX, containing the minutes of the Society and of the Mathematical Section for the year 1886.  8vo, pp. 113.  Price $0.75.

The Bulletins are sent to active members not in arrears for dues, and to corresponding societies.  They may be purchased of the Treasurer.

www.ingramcontent.com/pod-product-compliance
Lightning Source LLC
Chambersburg PA
CBHW051414200326
41520CB00023B/7226